Charles Van Norden

The Psychic Factor

An outline of psychology

Charles Van Norden

The Psychic Factor
An outline of psychology

ISBN/EAN: 9783337312824

Printed in Europe, USA, Canada, Australia, Japan

Cover: Foto ©berggeist007 / pixelio.de

More available books at **www.hansebooks.com**

THE PSYCHIC FACTOR

AN OUTLINE OF PSYCHOLOGY

BY

CHARLES VAN NORDEN, D. D., LL. D.

LATE PRESIDENT OF ELMIRA COLLEGE

NEW YORK

D. APPLETON AND COMPANY

1894

PREFACE.

THE purpose and spirit of this little book is strictly scientific. If any justification for its appearance be needed, the public will find ample in the unsettled condition of the metaphysical world, in the marvelous strides of biological and psychical discovery, and the utter demoralization of the old psychology. The Psychic Factor is not addressed to the populace, nor yet to original investigators, but to students. It is intended to embody the trustworthy results of safe thought in the realm of current psychology.

TABLE OF CONTENTS.

THE PSYCHIC FACTOR.

INTRODUCTION.

CHAPTER I.

THE SCIENCE DEFINED.

1. PSYCHOLOGY investigates mind. By mind we mean all psychic states, whether of intellection, feeling, or volition—in short, the psychic factor.

2. Recent psychology has been described as experimental and physiological, simply because practical rather than speculative—an attempt at exact science in a realm which hitherto has been proverbial for its vagueness, assumption, and contradictions.

3. All sciences have had their birth in regions of ignorance, and hence of superstition and of speculation. First came blank ignorance, after that superstitious interpretation, and then philosophic. speculation. Exact science began only with experimentation, and has proceeded by induction. Alchemy, at first a sorcerous attempt to convert stones into gold, only after ages of puzzled thought became chemistry. Suidas says that the Golden Fleece was simply a parchment on which was written this art of transmu-

tation. If so, it was then the first essay of an immense literature which only of late has possessed any value whatever. Star-gazing, an occupation for the eager and credulous, becoming astrology—which was but an instrument of fraud for charlatans—slowly evolved into exact astronomy. Ptolemaic cycles and epicycles, impossible though they were, at least prepared the way for Galileo, Copernicus, and Newton. And geology was conceived in ignorance and reared in a realm of legends, myths, and guesses. The fossil bones of giant beasts found in caves served to illustrate, for the devout, the story in Genesis of primeval antediluvian giants; while for the profane they buttressed credulity in thrilling tales of dragons and battles between beasts and men. Increase Mather—Cotton Mather's father—on finding ribs of the mastodon in Massachusetts, solemnly announced to the scientific world in England that he had discovered some traces of those infamous human monsters who brought on the flood. Within a century the common people of Germany have been known to grind up the bones of cave bears and hairy mammoths for potent medicine, believing them to be the remnants of griffins and unicorns. Petrified nautilus shells used to be called snake stones, and in popular belief they are still serpents beheaded and petrified by the prayer of sweet, timid St. Hilda. Less than one century ago Voltaire, sneering at the story of the deluge, was overwhelmed with the facts to which his attention was triumphantly called by Christian apologists, that petrified shells were to be found on the tops of mountains. The very lame defense was that these had not been dropped by the flood, but had fallen from the pockets of pilgrims returning from Palestine! Only yesterday did geology become exact.

Psychology has proved no exception to this rule. It has but just emerged from its threefold career through ignorance, superstition, and speculation. Only yesterday it was where physics stood in the days of Anaximander and Thales, in the stage of logical and descriptive analysis. And to-day the science is but in its infancy. Prof. James, of Harvard, says: " Psychology is to-day hardly more than physics was before Galileo, what chemistry was before Lavoisier. It is a mass of phenomenal description, gossip and myth, including however real material enough to justify one in the hope that its study may become worthy of the name of natural science at no distant day."

Not until the microscope established histology, and vivisection became the handmaid of surgery, and hypnotism opened a door into the realm of the subconscious, did an exact study of the psychic factor become feasible.

4. It does not, however, become us to underrate the value of the scholastic philosophy. Speculation in psychology has achieved real and great successes, owing to the possibility, in all ages, of introspection ; which is a very important department of scientific experimentation. The self-study of rare genius has not been fruitless. The results of the thinking of philosophers like Plato, Aristotle, Kant, and Hegel can never fail to compel respectful consideration. These results have proved of special value in the classifying of mental states. While psychologists have come to look upon the so-called faculties as only customary modes of mental behavior, the old-time discrimination of these

into groups for purposes of study has not been, nor is likely to be, improved upon.

5. But long ago the vein was well worked out; and on this line of effort nothing remains to do but to erect new structures of guesswork, and to batter down old ones just as well and just as ill founded.

6. The lackings of the speculative method are fourfold :

(1) Introspection begins and ends with the thinker himself; while the psychic factor is found in all living matter, and in every creature challenges study.

(2) In the thinker, introspection is limited to the present moment and to memory of a personal past. It can not investigate even its own evolutionary antecedents.

(3) It is utterly excluded from that vast realm of the subconscious, which forms the most interesting department in the psychology of to-day.

(4) It. is embarrassed by its own prejudices and delusions. It were easy to show that it constantly mistakes mediate knowledge for immediate, acquired knowledge for innate, and inherited for necessary.

7. Though it be in methods experimental, psychology needs must start out with a postulate. For there can be no explanation without something to explain, and all philosophy brings us to ultimates. Psychology runs back upon three ultimates—matter, life and mind. These it fails to analyze into simpler elements or to identify beyond a peradventure with one another. Speculation may, of course, devise hypotheses to analyze or to unify but all evidence fails. Hypotheses here have proved utterly barren. The endless battles of metaphysicians over these stubborn, elusive ele-

ments, in vain speculation, have been and can be only Walhalla conflicts of ghosts, that hew one another to pieces each day, on the morrow to renew the aimless strife joyfully and vainly.

8. We will note the five hypotheses that challenge our criticism at the present time, only to illustrate that folly of claiming to know too much which has always cursed psychology :

(1) Materialism. Matter embraces mind. Atoms and forces beget ideas, which are material entities. Consciousness is only a series of those mental materials. In short, matter has latent in it the promise and potency of mind. Mill, Spencer, Tyndall, and a host of English philosophers, and Comte and his disciples among the French, and Herbart and his followers among the Germans, have held this doctrine.

(2) Idealism. Mind embraces matter. Ideas beget atoms and forces. Material phenomena are only phases of consciousness. In short, mind has in it the promise and potency of matter. As Omar Khayyam in the Rubáiyát declares :

> " We are no other than a moving row
> Of magic shadow-shapes that come and go."

Hegel, Fichte, and Schelling stand prominent among idealists.

(3) Ideal realism. Matter is parallel with mind. Ideas and things, thought and being, are parallel one to the other. By some mysterious coarrangement (occasionalism according to Descartes, pre-established harmony according to Leibnitz) the two series accompany one another, mind and body perfectly attuned —soprano and bass—but without causal connection. Lotze seems to find refuge in such a scheme.

(4) Monism. Matter is mind. Ideas and things, thought and being, are identical. Consciousness is an aspect of certain material forces. Says Fechner, "What from an internal point of view seems to be your spirit, seems from an external point of view to be the bodily substratum of that spirit." If you stand on the inside of a hollow sphere you see only the concavity of its surface; if on the outside, only its convexity. Yet it is the same surface.

(5) Matter and mind are different in substance, but a causal connection exists between them. Atoms and forces on the one hand, and thought on the other! Body and soul, each substantial but of utterly different substance! The soul sways the body and the body affects the soul, there being a causal nexus. This is the popular, the traditional, and the simplest explanation. A very able defense of this view may be found in the closing chapters of Prof. Ladd's Physiological Psychology.

9. At first sight these hypotheses seem utterly diverse, and doubtless are advocated by men of widely different temperaments, beliefs, and tendencies. But the diversity at bottom is more seeming than real and is owing to our reading into them meanings derived from prejudice. If the first be true, and matter be capable of generating mind, then in matter must be dormant all the properties that can be shown to have ever inhered in the psychic factor, and it is no longer gross and inert but as truly spiritual as material. If, however, the second hypothesis be accurate, and if it be the function of mind to generate matter, why, thoughts are quite substantial and visions full of solidity, force, and point. And if matter and mind be parallel,

which is the worse or different for that fact? And if matter and mind be identical—only inside and outside of the same curved surface—what does this signify? On any of these hypotheses you have broached a possible fact, which is, however, utterly barren in practical bearings.

We naturally think of a materialist as a gross, wooden-headed, leaden-hearted thinker, but there is nothing to prevent in him the loftiest flights of poetry and spirituality. And we are tempted to judge the idealist as an insane skeptic; he may be and often is, however, a perfectly matter-of-fact person, given to hunger, athletics, and *bonhommie*. If you must have a hypothesis, choose one of these five, but do not deceive yourself with the idea that you have added to your stock of knowledge. So far as knowledge goes, there are three unresolvable ultimates—matter, life, and mind. These may be capable of resolution, they may in time be resolved, but as yet they are to all experimentation elements. And so the wag was quite right, who, when asked what mind was, replied, " No matter," and when asked what then matter was, answered, " Never mind." Well says Tyndall : " The problem of the connection of body and soul is as insolvable in its modern form as it was in the prehistoric ages." And Huxley : " How anything so remarkable as a state of consciousness comes about as a result of irritating nervous tissue is just as unaccountable as the appearance of the djinn when Aladdin rubbed his lamp." Yet both of these men are pronounced materialists.

10. Two affirmations we may safely make of the relations existing between the three ultimates, that emphasize the radical nature of the distinction :

(1) That particular molecules of matter may come and go in continuous interchange, without disturbing the vital or psychic processes in the least. This occurs in oxidation, secretion and excretion. In animal or plant, living matter is in perpetual flux.

(2) When mind and life depart in death the matter remains, so far as science can discover, chemically and physically the same.

These two facts throw discredit upon theories that confuse or identify the three ultimates.

11. Indeed, we shall find that while we push back the barriers of knowledge, beyond them lurk abysses we can not penetrate. Several things are to be remembered :

(1) We are dealing in this realm with forces infinitely more attenuated, subtile, and lively than we ourselves as a whole are. Thus, in vision we are dealing with light. Reflect that every second of vision, a cone of light for each luminous point viewed enters the eye at the rate of one hundred and eighty thousand miles. Gaze at a star. For every second of such vision a ray of light one hundred and eighty thousand miles long slides into the pupil. This ray is a vibration of ether and enters in the form of waves, and breaks upon the retina as the ocean upon the seashore in a sort of ethereal surf. In one second not less than five hundred billions of these light-waves dash into the eye to beat against the optic shore of nerves. And this shore of nerves is composed of one hundred million nerve elements to the square inch ; each of which receives a separate impulse, and all of which work in harmony. There is no end to such facts. We are subtler in detail than as a whole. As psychic beings

we master or are mastered by forces of inconceivable divisibility and subtility.

(2) Then, again, it is the fate of mind, when inquiring after causes and essences, to reach speedily the limits of knowledge; and to attempt to penetrate beyond is sheer folly. All the woes of metaphysicians have come from this foolhardiness. Ultimate facts will encounter us everywhere, marking the insurmountable barriers of thought. With them we must pause. Our final knowledge is and will probably ever be but a light shining in darkness.

(3) We are in a universe which bristles with problems that the human mind may at some future time solve, but which can not be successfully treated at present.

12. The purpose of experimental psychology, then, is not to find out all psychic facts, nor to find out any psychic fact to perfection, but simply to discover and arrange relative and derivative facts, to push out the barriers, to study mental methods, and to weigh the validity of mental operation.

13. It is therefore not a study of mental results but of mental processes—not what we see, but how we see; not what we think, but how we think; not what we feel, but how we feel. The results of mental action we have in various other sciences—astronomy, biology, chemistry, etc. Psychology tries to discover by what powers and methods we attain these results. It is the scientist turning from the world to study himself. It is a response to the everlasting "know thyself" of inquisitive philosophy.

14. This science is naturally and necessarily dominated by recent discoveries bearing upon the extreme

probability of an organic evolution. If man is the end of a series of growths allied genetically to lower minds, comparative psychology assumes first importance, and our study begins with the monad and finds a very good implement in analogy. In these lectures the evolutionary hypothesis is adopted as a working theory. It is not claimed as beyond peradventure proved, but only as best explaining the facts. Evolutionary language is used and the movement is by the paths of supposed evolutionary ascension.

15. The scope of this science is as far-reaching as the phenomena of life; for life never appears without mind as its correlative. Hence biology, the science of life, and psychology, the science of mind, are kindred in aims, and constantly cross one another's path. History recording the actions of men furnishes perpetual and diversified illustrations of psychic operation. As Herbart well claimed, "psychology shoots its roots into the sciences of life and blossoms in the historical sciences."

16. The future of psychology is very promising. Myers is justified in saying that "there will be no cause for surprise if, as time goes on, man's experiments on the world without should yield in interest and importance to his experiments upon himself. Inward the course of empire takes its way! . . . All that he has learned without himself has been but a means to the comprehension of that which was within."

CHAPTER II.

METHODS.

1. CHEMICAL and physical. Because mind is always associated with matter, and the laws of material composition and motion underlie the entire physiology of the senses and the organic functions. The study of the direct relations of mind and matter has been termed psycho-physics; it busies itself chiefly with the relation between the quality and intensity of stimulus and the quality and intensity of psychic reaction. Closely associated with this is psychometry, or the time-measurement of psychic reaction.

2. Biological. For mind is always associated with life, and we must study the living cell in its life history in order to investigate psychic processes in their simplest forms and lowest degrees. Here we need the microscope; for this reveals to us a new world of thought, feeling and action.

3. Anatomical and physiological. For mind at its best is associated with structures of great and significant complexity admitting of vivisection and dissection. Considerable knowledge of structure and function is now required by psychologists.

4. Pathological. Because mind in its various processes is associated with localities in structure. A study of disease often enables us to locate functions. This method, however, is exceedingly difficult and misleading, because of our ignorance of disease, and because of the interrelations of the entire nervous system. Valuable results will flow only from the most

patient, prolonged and unprejudiced observation and reflection.

5. Psychical. For mind is never found unassociated with other minds; and subtle interrelations are clearly discernible. Hence, hypnotic experimentation has of late years been much resorted to, as enabling one, the agent, in studying another, the sensitive, by suggestion and command to operate separately upon different sets of nerve centers. And hence, also, the value of studies in thought-transference, lucidity, and similar phenomena. These particular methods have come to group themselves together under the convenient heading, Psychical Research.

6. Introspective. Last and best. Mind's highest knowledge is self-knowledge. I am always *chez moi*— at home with myself. Introspection must begin and end, accompany and correct, all our devising, albeit with due regard to the necessary limitations.

CHAPTER III.

HISTORY AND BIBLIOGRAPHY.

1. EXPERIMENTAL psychology was rendered inevitable by the labors of Bacon and Leibnitz in the seventeenth century. Bain, Mill, Spencer, Taine, John Mueller, Weber, Fechner, and Lotze have been the prophets of its annunciation and exposition. Its present advocates are many, and they are the authorities in mental science.

2. Something of a division of labor, resulting from a varying direction of interest, occurs to-day among

psychical expositors. In England and America comparative psychology and psychical research are uppermost; in Germany, psycho-physics; and in France, pathological psychology.

3. Any one beginning in earnest the study is advised first to master a good history of philosophy, like Erdmann's or Ueberweg's, then Bain's The Senses and The Intellect, and The Emotions and The Will. After that, H. Spencer's Principles of Psychology. These simply to prepare the way for Lotze's Microcosm, Wundt's Physiologische Psychologie, and Ladd's Physiological Psychology. It will be well to add Janet's Automatisme Psychologique, if one read French and Binet on the Psychic Life of Micro-organisms. A great many articles in scientific magazines, pamphlets and books, of much value and intense interest, are now constantly appearing. Psychical research engages at the present time the rapt attention of many keen minds and the publications of the English Society of Psychical Research will repay careful perusal.

To keep in touch with the great leaders of what may be termed orthodox psychology one must be conversant with the works of such writers as McCosh, Sully, Baldwin, and Hoeffding.

PART I.

MIND IN GENERAL.

SECTION I.

THE PSYCHIC FACTOR CONSIDERED COMPARATIVELY.

CHAPTER I.

MIND IN PLANTS.

1. LIVING matter is always psychic. At first this statement may seem startling in its involvements, but it is simply a corollary not only of the theory of evolution, but as well of the unquestioned facts of development.

(1) We must infer it from the facts of development. Take our own human life-history, from ovum to maturity. The fertilized human egg is at first a single living cell, but directly, by division of the nucleus, it becomes an aggregate of cells. This aggregate takes shape as a sac or gastrula. Out of the gastrula arises a vertebrate creature of lowest notochord type; and this forms a spinal column, soon to appear as clearly a mammal. Last of all comes the human infant—child—youth—adult; and the series ends per-

chance in Plato, Shakespeare, or Tennyson. Now this progress is a close continuous unfolding of the original cell. There is no gap for mind to creep into during the movement. The original cell, simple and undivided, must at the start be viewed as psychic. If mind is to be found at last in Plato, we must presuppose it in the egg; if that egg be merely chemical and physical, why, so is Plato.

(2) We must infer the same from the evolution of intelligences in the organic world. There is an unbroken gradation of them, a series of ever-expanding numbers. There is no beginning place for mind anywhere in the evolutionary movement. Deny it at the bottom, and you must fail to get it all the way up. Claim it for the philosopher, and the claim runs down to the monad. Animal mind presupposes vegetal mind; and the mental rhythm of creation is dual. The logical result of denial will be complete skepticism, which ultimately must hold man himself as a mere machine—as, indeed, even so shrewd a thinker as Huxley has already boldly urged. Were there no evidence of intelligence in low forms, based on observation, we should need to infer it as at least a latent presence.

2. But we need not depend upon theoretical consideration; ample observation establishes the connection between mind and even the simplest life beyond peradventure. One may of course claim, that to infer intelligence from such action in low forms as would warrant the conclusion in human beings, is unjustifiable on the ground of the relative inferiority of the former. The difference, however, is rather of quantity than of quality. Purposive and ingenious conduct is

quite as trustworthy testimony in a micro-organism,
for instance, as in a domestic animal or a human
neighbor. The surety equals at least that of any observation of other intelligence than our own. And no
living form fails to furnish the desired evidence. In
this chapter will be cited in testimony those represent-
atives of life which, in common parlance, are named
plants, including the lowest protophytes.

3. The psychic difference between plants and animals is simply one of degree. Vegetals are feebly psychic, animals intensely so. To begin at the bottom,
let us consider the slime-molds, which are mere naked
masses of jellylike matter, chiefly protoplasm, that
ooze over decayed trunks of fallen forest trees or rot-
ting bark in tannery yards. Thiselton Dyer claims that
this inert, formless, functionless substance can be educated to the extent, at least, of learning to accept food
at first rejected. Moreover, the mass at some time in
its endless existence breaks up into swarm spores and
becomes active, with no little indication of psychosis.
All a-quiver, these swarm spores react upon stimulus,
and dart to and fro seeking food, with precision and
seeming prevision, until, this phase of life passing,
they fuse together to form new plasmodia. A very
large number of simple plants enjoy a similar period
of swarming and intense activity.

4. Scarcely higher—perhaps even lower—in the
scale are the microbes of putrefaction, fermentation,
and disease. P. F. Frankland, in a popular lecture on
these forms, claims for them individuality and capacity
for education. He says: " In fact, experimenting with
micro-organisms partakes rather of the nature of legis-
lating for a community than of directing the inani-

mate energies of chemical molecules. Thus frequently
the past history of a group of micro-organisms has to
be taken into account when dealing with them; for
their tendencies may have become greatly modified by
the experiences of their ancestors." However this
may be, it is now known that bacteria possess an oxygen sense, by which they detect the presence of oxygen
at a distance and are able to seek it. Moreover, they
can gauge the quantity of this gas, and if it prove too
intense they flee it. Engelmann claims that bacteria
can detect one trillionth of a milligramme of oxygen,
or, in other words, a solitary molecule.

5. The Desmids, a pretty order of green plants,
each of but one cell, possess a sunshine sense. We
discover no pigment spots, but without failure they
find the sunny side of the tumbler in which they are
imprisoned, in order to expose to the sunbeams their
chlorophyl and work their simple machinery of nutritive assimilation. They distinguish light from darkness, and in the light find the sunbeam.

6. In colonies of *Pandorina*, a higher form, some of
the cells possess pigment spots that serve as rude eyes
—not, of course, for vision, but simply to sense the
sunbeam. While the Desmids have only a light sense,
Pandorina has a light sense organ.

7. Many colonies of one-cell plants show a sort of
aggregate intelligence. These are veritable confederacies, not organically one, but dominated by a common
purpose and united in common movements. We may
still have swarmspores or zoöspores, and a motile
period and mutual attraction and fusion; but there is
added combination for security, motion and nutrition.
Oscillatoria is a case in point; it is simply a cylindrical

filament of cells, each shaped like a pill box and piled
end on end. Its confederate action is merely a sway-
ing of the filament to and fro in rhythmic oscilla-
tion. The psychic harmony seems based upon protoplasmic contact.

8. More surprising are those combinations of cells
which form not so much confederacies as true unions,
with organic interharmony and public functions, also,
doubtless, based on protoplasmic contact. These are
often spoken of as (vegetal) "persons." A fine example is *Volvox*—a beautiful sphere of cells, the whole no
larger than the point of a pin. There are twelve thousand individuals, each with two protoplasmic lashes
called cilia. The twenty-four thousand lashes all wave
rhythmically, and the community rolls through the
water with perfect unity of purpose, protoplasmic contact of individuals enabling the whole to move as one
creature. Moreover, this confederacy owns common
duties, and a true division of labor subserves the common end. Some of the cells in partnership develop
spermatozoids, others become oöspheres, and, through
the combination of the two, new colonies come into
being. These baby volvoces are protected during expansion within the hollow globe, until set free by the
death of the parent colony. The older sphere rolls
slowly across the field of the microscope, within it revolving the spheres of the future, like the vision of the
prophet, "a wheel within a wheel." *Volvox* is clearly
swayed by protoplasmic unity, and that sway is psychic.

9. Tissue plants only emphasize the same unity of
plan and action, controlled by forces other than chemical and physical. And we no longer wonder over the

growth of a fern from spore to frond, or of a palm from cocoanut to plume, or of an oak from acorn to leafy crown. In all these cases one cell multiplies into many, and these come to exhibit, in an orderly way, the nicest specialization of parts and division of labor and concert of action, all on a plan foreordained in the fertilized ovum, and in every case peculiar; and each form, thus highly organized, lives and dies as one person. Many tissue plants seem much less psychically active than those which are microscopic, but the psychic activity is only latent; for during the period of fertilization both sperm and germ cells become amœboid and assume a motory existence. Often the sperm cells are true spermatozoids, and not only very active, but with display of instinct. They respond to stimulus, and actively seek the germ cell of their own species, which they recognize and approach. It is known that the spermatozoids of ferns are attracted to the corresponding archegonia of the prothallus by malic acid, which is secreted in the latter to attract and guide them. Some tissue plants, however, are intensely psychic at all times. Some are exquisitely sensitive, like the well-known mimosa, and faint at a touch. These barely fail of a true nervous system, their protoplasm being almost as sensitive as nerve matter. Others are carnivorous and entrap animalcules, insects, etc., to consume them as food; which is true of the bladderwort, pitcher plant, Venus's flytrap, sundew, and water pitcher.

10. These results are secured somehow by the actual contact of all the protoplasms combined. Minute filaments of living matter, through even thick walls of cellulose, connect not only neighboring cell plasms,

but even the chromatin of neighboring nuclei. A network of living matter explains the harmony of plan and action.

11. All tissue plants exhibit what is called geotropism—that is, their stems grow upward and their roots downward, or they somehow arrange themselves with reference to the force of gravity. Moreover, some runners and some rhizomes grow horizontally. Some flowers face and follow the sun. Some tendrils reach out for support and about it twine themselves. Some plants sleep at night, allowing their tissues to become flaccid and to droop. All these movements have hitherto been explained by the best botanists as mechanical. A closer study of the facts' now casts doubt upon this far-fetched solution; and Francis Darwin does not hesitate to ascribe plant movements, like those of animals, to irritability aroused by stimulus.

CHAPTER II.

MIND IN ANIMALS.

1. In animals, protoplasm assumes more active habits, mind predominating over matter. The creature on its entire periphery exhibits sensitiveness to stimulus and responds in appropriate action, showing feeling, volition, and judgment. It is active, and hence voracious; predatory, and hence ferocious.

2. Animals of one cell, or protozoans, invariably put forth motory organs, as do plants in their motile states. In the rudest, these organs are extemporized out of the body substance—mere protrusions or false feet (pseudo-

pods). In the higher they become quite permanent lashes (*cilia*) or whips (*flagella*). The lashes are short and delicate, and often form a light fringe. The whips are long and powerful. Both are composed of two united filaments of protoplasm, the one contractile and the other elastic. Hence they can bend and rebound. Of the whips there are two kinds, the one trailing behind, the other extending in advance. The forward whip is used, as a boy uses his right arm when swimming on his right side, to draw the body along. The hinder whip serves like a tadpole's tail, to propel. These members secure and guide the animal's movement when in search for food, or, by creating a vortex in the water, they bring in the food from a distance, while the creature remains still.

3. Protozoans have the rudiments of senses. Thus all have the power of touch seemingly on the entire periphery. Probably also they have what is called " general sense."

4. The light sense is well developed in some instances, localized in pigment spots, which give at least a perception of the distinction between light and darkness, sunshine and shadow. *Euglena*, a pretty green infusory, has a pigment spot of bright red, which is sensitive to light and enables it to seek the sunbeam. As this is one of the chlorophyl animalcules with a vegetable habit, of course light is necessary to its existence, and the organ of great practical value.

5. Even the sense of hearing has been claimed, at least for one beautiful ciliated infusory—*Loxodes Rostrum*—which exhibits along the back a row of small organs supposed to be of the general nature of auditory sacs.

6. They may or may not have smell and taste; but the presumption is in favor of at least smell, in view of their skill in finding and selecting appropriate food. They certainly show in this regard likes and dislikes; and then some are herbivorous and some carnivorous.

7. That they have the rudiments of sensation and perception follows from the possession and use of senses.

8. That they have the power of judgment on a low scale is also to be inferred from the purposive nature of their activity, and the instinct, experience and skill evinced in their methods of life—as in the pursuit of food; some go in quest of it, some draw it to them by creating little vortices with their fringe of lashes. This is readily studied in the process of decomposition. After the bacteria and other simplest forms have caused extensive decay of the tissue, a creature appears on the scene armed with a long, rigid, anterior flagellum terminating in a hook, and a posterior flexible flagellum. It anchors itself by its trail and then successively coiling and uncoiling the flagellum, darts up and down on the decaying substance. This infusory is succeeded by a group that hurl themselves on the putrid tissue and hammer it to bits. Finally, a gleaner appears and devours the scraps. Each form comes at the right time, recognizes the presence of appropriate nutriment, treats the food in a skillful manner, ingeniously satisfies its needs, and sustains life until changed conditions force out of existence everything but its dormant germs.

9. Or take their methods of attack, as evincing not only thought but also volition and feeling. According to Stein, the *Bodo Caudatus* combines in companies

of ten, twenty, forty, for purposes of attack. Like wolves, these little flagellates will throw themselves upon animalcules a hundred times larger, worry them, tear them to pieces and devour the huge prey piecemeal. Many hunter infusories are supplied with trichocysts, by means of which they wound, stun, and disable their quarry; while other more peaceable animalcules are armed with the same weapons to be used only in defense. The trichocysts are sharp filaments, poisonous, like the stings of nettles, with which the parts adjacent to the mouth supply themselves by some internal method of manufacture. These serve as darts and are shot out by some simple mechanism with sufficient force to pierce. The attacked animal, wounded, is paralyzed, no longer tries to escape and is easily devoured.

10. The protozoans are in some cases, as with *Vorticella*, distinctly male and female, and these share in the usual psychic phenomena of sex. It is said they make love and indulge in coquetries, the male seeking and the female exercising choice.

11. Communal instincts show themselves in the groupings of protozoans into colonies. Thus *Vorticella* may in the same genus occur with one species in separate individuals, and with another in a compound arrangement; and in the compound arrangement there is not only individual but communal sensibility.

12. Metazoans are protozoans complicated by development and evolution; for every metazoan begins its life-history as a protozoan, and must have arisen in the combination of a number of single cells of common ancestry. The simplest conceivable form of metazoan is that of a double sheet of cells rounded into a

pouch. The inner cells become nutritive, the outer serve for sensation; and the whole creature is a true federal union descended from one egg cell. From this simple scheme, by growth and specialization, all elaborations of animal shape and function are possible.

13. Protozoans show tendency in many genera to become metazoan. Thus in the genus *Zoöthamnium*, while some species are mere colonies (*simplex, nutans*), others are true federations (*arbuscula, alternans*).

14. The intelligence of metazoans is now so generally admitted by scientists that we need not carry the argument further. Presuming that the psychic factor is here acknowledged, it seems important rather to expend energy upon its elaborate methods of assertion.

CHAPTER III.

THE NERVOUS SYSTEM.

1. THE growing complexity of metazoans, as we ascend the scale of animal existence, ere we proceed far, necessitates the specializing of certain cells to regulate the others. Protoplasmic unity of all the cells fails to meet the demands that arise for nice coordination and vigorous purposive action. The psychic factor itself, now that many individual cells are cumbered with special functions, demands an organ. The little union has become so complex in its public duties and interrelations that there needs not only consensus and harmony, but authoritative government.

2. For this, nerve cells are set apart. These, how-

ever, seem to have no properties unpossessed by protoplasm in general ; indeed, they are only commonplace
individuals specialized to perform certain difficult
functions—just as a president, a senator, or a gov-
ernor is, after all, only an American citizen, in no wise
different from farmers, traders and mechanics, except
in the fact that he is set to govern all the rest. Nerve
cells are of various shapes—spheroidal, ovoidal, trian-
gular, etc., and have nucleus and nucleolus.

3. In its rudest form the nerve cell has no attach-
ment, and exerts a direct control over adjacent individuals by mere protoplasmic contact, as in *Hydra*.
In general, however, it exhibits processes—one, two, or
many—which are called nerves, and themselves, probably, are elongated cells end to end.

4. At its simplest a nerve is a protoplasmic fibril
and nothing more. In animals well innervated it becomes a fiber, composed of many such fibrils, or it may
be a bundle of such fibers. In the higher vertebrates,
where dense masses of nerve cells and infinite complexity of interrelations require it, the fibers are often
united in skeins and the skeins in cables. A vast
number of fibers may be in one so-called nerve.
Thus, in the motor nerve of the human tongue are
full five thousand fibers and many hundreds of thousands of fibrils. In what we call the optic nerve—in
reality an optic cable—are no less than one hundred
thousand fibers and many millions of fibrils.

5. In the vertebrates (except in the sympathetic
system) nerves are insulated by two layers of non-
conducting material up to near their terminations
and are divided lengthwise by nodes somewhat like
a cane or rattan stalk. All human nerves are supplied

3

with what seem to be re-enforcing cells—relay batteries—here and there along the line.

6. The function of nerves is simply to convey impulses received either at the center or on the periphery. Hence they connect a nerve cell with another such cell or with an end organ. In passing through intermediary nerve cells the composing fibrils spread out from pole to pole against the inner surface of the cell wall, so as to inclose the cell protoplasm. Hence nerves are said to be efferent or afferent, according as they conduct energy from the centers to the periphery or the contrary. This division of labor is not based ·upon any essential difference in composition and it is likely that both kinds are able to convey impulses in either direction.

7. The energy, like other forms of motion, is a transformation in this case of chemical affinity if from within, or of the impinging energy if from without. It is one motion converted into another, probably by the nerve cells, but become peculiar to itself, and quite different in physical properties from light, or heat, or electricity.

In man a sensory impulse has been calculated to travel at a speed of from one hundred to three hundred feet a second. Moreover, the number of impulses that may thus travel during one second is very great. By stimulation with the wires of a telephone, it has been shown by D'Arsonville that a nerve can transmit upward of five thousand vibrations per second, and that the wave-forms may be so perfect that the complex electrical waves produced in the telephone by the vowel sounds can be reproduced in the sound of a muscle after having been translated into nerve vibrations and transmitted along a nerve.

8. It is the nerve cell that stores up complex and unstable material derived from food, and by exploding this releases the needed energy. Morgan likens this complex material to a building of wooden blocks erected by a child, which, the more elaborate it becomes, the more unstable it is, until a jar or a touch shatters the edifice, liberating the stored-up energy of position acquired by the blocks in building. We may also liken a nerve cell to a pistol loaded—a touch on the trigger discharges it. Or it is a galvanic battery, and the costly stored-up material is the zinc, to be consumed when the current is closed and the imprisoned energy is released.

The shock may be very violent, as in the case of Sulla, who is said to have died of an explosion of wrath ; or of Leo X, who fell a victim to an outburst of joy. It has been ascertained that bee drones perish in the act of sexual intercourse, slain by the shock of passion, and not, as formerly asserted, by ruthless action of the queen.

9. A nervous system is a series of two or more nerve cells with appropriate connections ; it may be very simple or it may be very intricate. A typical system is a solitary psychic cell connected by two protoplasmic filaments with a sensory and with a motor surface, but any variation on this is possible.

As nerve systems become elaborate, they themselves need central control, hence centers simply for co-ordination of such centers; these may be termed co-ordinating organs.

10. Nerve cells are capable of three kinds of action :

(1) Automatic. The word explains itself, and a good illustration will be found in the cells that control the beating of the heart. A frog's heart after excision will yet throb some time, because the ganglia that control the action accompany it.

(2) Reflex—that is, in response to stimulus come in on some nerve. A pinch of snuff, and the resulting sneeze, will serve for illustration.

(3) Inhibitory. Where one nerve cell prevents or restrains action in another. Stimulate the vagus, and the action of the heart is arrested. Atropine paralyzes this inhibitory influence, muscarine stimulates it. If from any cause the smaller arterial vessels become constricted, and the heart in forcing blood through them be required to work with greater effort, and so in danger of exhaustion, the depressor which connects the heart with the vasomotor center inhibits or depresses this center and obliges it to dilate the vessels, and so remove the cause of the embarrassment.

The entire nervous system is such a marvel of complex harmony, capable of intelligent supervision of its own operations at every point.

11. Four laws govern the action of nerve cells and systems :

(1) Of specialization—that is, of specific function for every element. Of course, in simple forms, the function may be much more general than in the more complex. The law becomes more and more emphatic as we ascend the scale, until in man specialization is carried to the extreme.

(2) The law of habit. Energy here, as everywhere, follows the lines of least resistance. Psychic action carves out physical channels in the process of time

and flows in them as a matter of course. As Carpenter puts it, the nerve system "grows to" the modes in which it has been exercised; or, to vary the illustration, nerve matter, like paper, folds most easily in the old wrinkles. Indeed, it is claimed by evolutionists that nerve fibrils owe their origin, in the gradual development of systems, to this very law; lines of motion following paths of least resistance occasion neural trails. The trails become fibrils or filaments of cells end to end, specialized as neural highways.

(3) The law of duration. Nerve reactions are far from instantaneous, and can easily be measured. It takes time for a stimulus at the periphery to reach the center, time for the center to receive the impulse and to respond, time for an execution of the motory result. The moment elapsing between a stimulus and its result is called "reaction time," and is different for different individuals and for differing moods of the same individual. Much labor has been put forth in its careful estimation, especially in Germany; but the results so far have been disappointing in the matter of important discovery.

(4) Conservation of energy. The nerve systems exert no force not derived; their motions are previous motions converted. Their explosions are exhaustive and their wasted energies must be redintegrated. Hence all complicated centers are abundantly supplied with nutriment. The brain, during action, is suffused with nourishing blood. The immediate concomitant of an effort at hard thought or intense feeling, or vigorous willing, is a rush of blood to the ganglia in use; and so if one be hungry, weary, or anæmic, the effort is likely to prove feeble. Centers

can not generate force of themselves, but can only use
such energy as the nutritive apparatus supplies. Hence,
for vigorous psychosis, the need for substantial food
well digested.

CHAPTER IV.

A SURVEY OF NERVE SYSTEMS IN THE ORDER OF COMPLEXITY.

1. PROTOZOANS, as we have seen, have no nerve
systems, but their single cell is itself the prototype of
the nerve cell, possessing, at least in germ, all the
qualities of automatic and sensitive life which the
latter attains.

2. The first appearance of cells specialized for sensory-motor control is found in *Hydra*, one of the simplest of the metazoans, which is yet complex enough
to require central control; it has nutritive, sensitive,
combative and generative cells, and in addition a few
psychic; these however without processes. Hydraform metazoans go no further on this line of evolution.

3. Medusaform metazoans, which are simply an
elaboration of the hydraform, develop a more extensive nerve apparatus. Thus the medusæ of *Bougain-
villia* present a central and a peripheral system, the
former a double ring of nerve cells, the latter scattered
nerve cells, and both connected with pigment spots,
muscle fibers and sensitive ectoderm by filaments.
The higher medusæ perfect this simple arrangement
and add olfactory tracts to rude eyes (or ears). Cut a
hydra into bits and each part will restore a whole
creature. The specialization is not complete enough

to cripple the generative and formative independence
of each cell. But if you slice off the nerve ring of a
medusa, what remains of the animal is thereby paralyzed, and will soon die. Specialization cripples cell
independence and though reflex action on stimulation
of peripheral centres occurs, automatism ceases.

4. The next higher order of radiated metazoans
(the *Actinozoa*) improve somewhat on medusaform.
Coral polyps, sea-pens, anemones and the like possess
in some cases rudimentary eyes and ears, with corresponding co-ordinative organs, but with similar limitations.

5. The *Echinodermata*—sea urchins, starfishes,
crinoids, etc.—also group the nerve centers in a ring
about the mouth. There are ganglia for each ray,
connected together by filaments, and from each gan-
glion there are radiating nerves. Thus the creature has
as many little brains as rays; these, however, though
acting in concert, must not be conceived of as entirely
dependent. Cut out one segment of a starfish and it
will thrive very well, under its own local control. The
brains are harmonized, but not co-ordinated by any supreme center. Hence the radiated animals have not
attained any very high grade of intelligence.

6. A more hopeful plan has proved that of the articulated animals, whose bodies are made up of segments—worms, centipeds, etc.; each segment is like
the previous one and each has a very simple nerve
system connected with all the others by double filaments. The chain of cells is ventral, and a pair of
ganglia for each segment. Here the system is of the
simplest, but it is reduplicated. And now we are in
the line of royal succession.

7. An improvement on this appears in certain of
those worms which have cephalic ganglia, the twin
nerves rising and dividing to allow the œsophagus to
pass through, and ending above in one or more highly
psychic cells. In this arrangement segments are still
independently active. Cut off the head of a centiped
while walking, and its body will continue to move onward; cut the body into three or four parts, and the
same result will obtain. Let the headless end strike
an insurmountable obstacle and, while of course the
motion ceases, the legs will continue to strive in
attempted propulsion. The cephalic ganglia supply
only needed general direction; when the other nerve
centers are destroyed, these fail of instruction from
headquarters and are like a company of soldiers in
battle whose officers have been killed.

8. Among the insects this scheme becomes much
more elaborate, and the cephalic ganglia much more
numerous and complicated, with an immense stride in
psychic energy. Thus the ants possess a real brain,
large in proportion to their size and bearing some
faint resemblance to that of the vertebrates; and they
seem to have carried the type they represent to nearly
its full ideal development. Still, we are hardly prepared by the visible anatomy of an ant's head for the
astounding unfoldings of mind manifest in its life's
history. Sir John Lubbock, who has made a lifelong
study of these insects, declares that they rank in intelligence next to man. He has discovered that they
possess character, and are some timid and some bold,
some born to lead and some to follow, some thievish,
some greedy, some phlegmatic. If they fail of a language, they at least transfer intelligence readily by

crossing antennæ. This has enabled the organization of quite complicated social systems, with the virtues of public spirit, neighborliness and patriotism carried to the utmost limit of personal abandon. In some species distinct classes have evolved—warriors, workers and slaves. Communal industries flourish; many kinds keep aphides and regularly "milk" them for their honey; *Lasius flavus* preserves in its nest during the winter eggs of these lice, hatches out young by a sort of incubator process, and in the spring stocks appropriate trees with them, quite as men breed cattle, stall them and then send them out to pasture. In western Texas, *Myrmica barbata* clears a tract of ground about four feet square around its city, and from this garden spot all plants are rooted up, and all stones and rubbish removed; a variety of millet is now sown, weeds that spring up extirpated and marauding insects warned off; when mature, the crop is reaped and stored away in granaries within the nest, for winter consumption. The parasol or leaf-cutting ants of Trinidad plant a fungus garden and nourish themselves on the proceeds of their labor. Many species keep slaves, who do all the hard and dirty work; these are seized in their homes, in the larva or pupa form, during great forays and often amid bloody battle, and are brought in the jaws of their captors to their new abodes, where they are taught to serve and wait; and there are good and there are bad masters. The foraging ants of South America make incursions, sometimes in dense Macedonian phalanx, sometimes in light detached columns; they send out scouts, survey routes, convey to one another information and form camps, quite as though human beings; every few days they arrange new camps, like the cruel robbers they are, move to and fro over the country, according to the exigencies of their predatory existence. The ingenuity exercised by these formidable barbarians in overcoming obstacles encountered on the march is said to be astounding. Thus, in crossing a crumbling slope, which was gradually disintegrating under the passage of the army, a portion of the band, by adhering to each other, formed a solid pathway over which the others passed safely; a twig formed a bridge across a small rill, but this proving insufficient for the transit of the host, it was widened by ants clinging to each side of the twig.

There seems no end to these wonders, nor any good reason for not comparing such intelligence favorably with that of the Australian bushmen, the Veddahs of Ceylon, or the pygmies found by Stanley in the African forests. From the psychological standpoint these facts are to the highest degree significant.

9. The nerve system of articulated animals is improved upon by a fusion of the chain of ganglia into a continuous mass or "cord." This appears in the lowest of the vertebrates, and is supported by a flexible fibro-vascular rod called a notochord. Amphioxus, a stupid, senseless little creature, and certain genera of fishes, are so equipped; and the embryos of all higher vertebrates pass through this notochord stage. But the latter in time develop a true vertebral column embracing in its bony canal the spinal cord.

10. The spinal cord is found to enlarge at points where its resources are severely taxed by limbs or sensory organs. In *Amphioxus*, which has no limbs, and few if any sense organs, and the lowest fishes with

cartilaginous skeletons and a uniform wormlike body, the cord presents the same general appearance at every point; but in fishes endowed with powerful fins and some intelligence there are corresponding local enlargements. These great and continuous bodies of nerve matter are unaccompanied by much psychic energy, as they serve simply to co-ordinate correspondingly large masses of end organ and muscle.

11. The enlargement of the cephalic portion of the cord first in fishes produces a true brain. A brain is thus a cephalic collection of specialized ganglia; its appearance signalizes the presence of energetic psychic activity. In its lowest forms a brain is composed of very simple sensory and motor ganglia, and the vertebrate so endowed is far below bees and ants. As cephalic ganglia become complicated, they themselves need co-ordination—hence centers for brain co-ordination. Over the sensorium rises a cerebellum, and over that a cerebrum.

12. In all but the lowest of the fishes we have a distinctly marked cerebellum, double optic and olfactory lobes, with two diminutive cerebral hemispheres. The development of the cerebrum now significantly marks the progress of intelligence. In *Amphibia* the hemispheres are relatively larger than in fishes. In reptiles they push backward, in birds both forward and backward. Anterior lobes of the cerebrum only are found in egg-laying vertebrates. First in placental mammals appears that great body of connecting fibers uniting the hemispheres and called the corpus callosum. Rodents give us the earliest indication of middle lobes distinct from the anterior. Monkeys develop posterior lobes, and these the anthropoid apes

emphasize. Only in the more elaborate mammals do these fold in and form convolutions ; the hemispheres now divided and convoluted quite cover the cerebellum and the medulla, and a forehead occasioned by en-larged masses of ganglia may be perceived. At last we have the human brain with a cerebrum whose cortex is folded into fissures and crevasses, until its surface is doubled, and with correlated ganglia aggregating full six hundred million nerve cells, and probably a much larger number of nerve fibers.

CHAPTER V.

GENERAL REFLECTIONS UPON THE PSYCHIC FACTOR COMPARATIVELY VIEWED.

1. LIVING matter can always be described in lan-guage of mind.

2. The ascent—from simple to complex—is marked by an ever-increasing specialization of cells for psychic functions, enlarged function in general indicated by enlarged ganglia. Animals that depend much upon vision are sure to display extensive optic lobes, those living by scent great olfactory tracts. Birds that use wings for flight, in the corresponding vertebral gan-glia show significant increase in size, while those that depend exclusively upon the legs for locomotion indi-cate this by the swelling of the spinal cord lower down.

3. As a rule, the more nerve centers the more men-tal functions. Each specialization means a corre-sponding dexterity in the psychic factor. Co-ordi-

nating lobes counterbalance diversity and restore personal unity to organisms that would be otherwise overspecialized; hence these indicate high intelligence.

4. It must be remembered, however, that increase in the size of ganglia may be owing to mere enlargement of bulk; in this case massiveness is not significant. Both the elephant and the whale in brain weight excel man; but this size is only the correlative of bodily bigness. It is the relative weight of nerve matter that signifies.

5. And even the relative weight is deceptive, without regard to quality. Many apes possess more brain than man in proportion to avoirdupois, the difference being in quality. The brain of the ant is, as we have seen, an instance of the remarkable possibilities of even minute particles of nerve matter. We may safely say that the amount, complexity and quality of mind, in a general way, correspond with the amount, complexity and quality of nerve matter. It is sound psychology to speak of a "brainy man," and Henry Ward Beecher was right when he declared that the world had always been swayed by men of "big heads and big bellies."

6. Added ganglia often re-enforce those previously existing, by contributing higher potentialities. At least four or five of the basal lobes of the human brain are united in the work of co-ordinating muscular movements with sensation. All the great nerves of sense spring from two or more roots, imbedded often in quite different soils; thus the optic nerve arises by different roots from the optic thalamus, corpora quadrigemina and geniculata, to say nothing of adventi-

tious roots connecting one tract with the other. This means that the apparatus for innervating eyes in the lower animals, in the higher is re-enforced with new potentialities. It is very clear that eyes, ears, etc., are far more varied in endowment with mammals than with articulates, radiates and mollusks.

7. Nerve masses which were once centers of consciousness, as more elaborate organs appear, work automatically or in a merely reflex activity. *Amphioxus* does all its thinking with its spinal cord; but vertebrates that have risen to the dignity of a brain, use the spinal cord only for reflex, conductory, and automatic work. Consciousness with each step upward becomes more comprehensive and intense, rising to higher outlooks upon the universe and more subtle and complex intellection, but seemingly withdrawing itself from regulation of the lowly functions of mere bodily existence, which fall to the realm of all but the highest centers.

We may in ourselves observe this withdrawal in constant operation. We learn in childhood to walk, with much painful education of reluctant nerve centers; but in boyhood already walking has ceased to be a matter of conscious regulation. One is taught to ride a bicycle with many woful episodes of inexperience, as a necessary concomitant of the intensely conscious process; in time the bicycle becomes part of the rider, and he now recognizes passing friends, enjoys the scenery and muses undisturbed as he skims along. The same is true of reading, singing, pianoplaying, and even of preaching and praying; the centers run themselves. In these cases consciousness has by no means withdrawn wholly, but it evidently tends to do so; give it ages of evolution, and it might do so entirely.

8. This withdrawal in no wise interferes with the automatic and reflex performance of duty on the part of lower centers. We have seen that to cut off a centiped's head does no more than remove control; the after ganglia operate normally. A frog from whom the cerebral lobes have been removed will swim, leap, crawl and croak; it is sensitive to light, but is stupid and listless, its life but a dream. Even a rabbit so operated on will stand, run and leap, start, tremble, cry if pinched and seek the light; but it is torpid, its consciousness that of sleep. A bird thus maimed will pick up food, drink, fly, clean its feathers, avoid obstacles and start at sharp sounds or flashes of light, but is dull and sleepy; nay, it is asleep, only the lower ganglia present and active.

Human beings have in many instances lost large masses of brain matter without serious impairment of faculties. Lallemand narrates the case of a person of average intelligence in whose cerebrum the right hemisphere was found after death to have been filled with only a serous fluid. Boyer tells of an epileptic child of usual brightness whose entire temporal lobe on the left side was found to have been destroyed. A premature discharge of blasting powder on a certain occasion sent a crowbar through the head of a young American; entering at the left angle of the jaw and passing through the top of the head, it was picked up some distance off smeared with blood and brains. The stunned youth recovered in a few minutes, ascended a flight of stairs, gave an intelligible account of the loss to a surgeon and continued to live for over twelve

years, with no impairment of his sensory or motory
powers. Human infants born without other brain
than the medulla have been known to live for hours,
crying and sucking.

9. Ganglia may act vicariously. When a nerve
center is destroyed a neighboring center often can and
sometimes does assume its *rôle ;* for it must be remembered that nerve cells, after all, are only protoplasmic
elements specialized, and that they preserve somewhat
of the power of general adaptation to environment and
stimulus. Just as a factory hand, who all his life has
devoted himself to some one little operation in the
making of shoes, watches or sewing machines, yet can,
if necessary—though doubtless crippled by such monotonous activity—do many other things.

10. The evolution of mind displays a marvelous
unity in diversity. At the beginning of the individual
life, and at bottom of the psychic scale, we have
nothing greater than the living cell ; and at the end
of the individual life and at the top of the psychic
scale, there is nothing greater than the living cell ;
there is no break in unity, and only growing diversity
with three great leaps.

The three leaps are: (1) The appearance of protoplasm in form of cells. This made structure possible. (2) The specializing of cells. This made function possible. (3) The co-ordinating of functions.
This made all degrees of mental attainment possible.

11. Mark the progress in its results. First single
cells with a psychic factor and conscious of their own
simple activities; then colonies of such cells pervaded
by a fellow - feeling; then communities, federally
united and with a communal consciousness, the indi-

vidual cell-mind now tending to work automatically; then communities in which some cells are set apart to feel, think and will for all the rest—in short, with nothing less than veritable government. Individual cells with the rise are more and more automatic, and their consciousness retreats into the background. The final outcome and greatest triumph of Nature is in the evolution of large masses of nerve matter for very elaborate psychosis, with consciousness covering the play of only the highest centers, the lower groups acting automatically.

Crowning the whole, man !

12. We have no evidence that Nature's reservoir is exhausted in man, even on the lines of neural development. Who knows what further possibilities of brain development and complexity may not exist? Who shall say what future evolution may not do for man's present brain ? Who can tell what other and better endowed creatures may not somewhere, or even here, arise ?

CHAPTER VI.

CONSCIOUSNESS IN GENERAL.

1. CONSCIOUSNESS is an ultimate fact, and therefore does not admit of definition. Every one knows what it is until asked to tell. It is not a mere name for a series of mental states, for these suppose its presence; it is not any particular psychic operation, because it reviews all psychic operations. It is a recognition by mind of its mental states, an awareness of what is going on within, and thus mentality in its last analysis.

2. The organ of consciousness is primarily living matter. There seems no good reason for denying even to the lowest forms of life some at least dim and shadowy awareness of their psychic acts. All that we have thus far said emphasizes the justice of this claim. When nerve centers appeared, these doubtless functioned as the exclusive organs of consciousness, which had withdrawn from commonplace cells; and when nerve systems were organized, the last-formed became the seats of conscious existence. In man the organ of consciousness, with the greatest probability and accord-

ing to nearly all competent thinkers, is the cortex of the cerebrum.

3. The basic fact of consciousness is change. Possibly without change we might have knowledge, but we should hardly be aware that we knew: consciousness would become *nirvana*, and practically extinct. The awareness is at least kept alert by change, and in all conscious life change is incessant.

Moreover, change has a physiological necessity; for, as our nerve centers are constructed, action always involves exhaustion, and persistent use destruction. Healthful activity of the brain constantly shifts the burden from cell to cell, from one center to another. Ribot, describing the tumultuous stream of thought, calls it " an irradiation in various directions and through various strata—a mobile aggregate, which is being incessantly formed, unformed, and reformed."

4. Hence, consciousness involves a time consideration. It is a constant *now*. It is aware of memories, but not of those past occurrences and operations themselves which are remembered ; it is aware of anticipations, but not of those future occurrences and operations themselves which are impending; it is aware only of present mental states.

The *now* of consciousness is not a point but a period of time—very brief, but of sensible duration, with a fading indistinctness behind and a brightening indistinctness before. The length of this period varies from six to twelve seconds.

5. Consciousness involves a discrimination between an ego, or self, and a non-ego, or not-self—that is, between a conscious subject and an object of which the subject is aware. This has been denied, on the ground

that babes are not supposed to make any such discrimination, and that they discover the self after a while. The objection, however, is only a surmise, and not an overwise one. There seems no good reason to deny that an infant, even at birth, may have an awareness of itself at least dim and empty enough to correspond with the void and shadowy nature of its consciousness at that time. It would not be in keeping with our purpose to dwell upon the idea of self in a metaphysical spirit, and it is here simply postulated without speculation as another of our ultimates, incapable in its last analysis of definition.

When consciousness is busy with its own states, viewing them as its own, we name the operation self-consciousness.

6. Consciousness has two functions of supreme importance—attention, and the enchaining or grouping of mental states. Without these wonderful gifts the so-called faculties would each one be quite useless. The first of these two functions is consciousness intensely aware, and the other is consciousness aware of the relation between its objects.

CHAPTER VII.

ATTENTION.

1. ATTENTION is a temporary arrest of psychic change, a fixation of consciousness. If we picture the latter as the mind's eye, the former will be the "yellow spot" of clearest vision.

2. The compass of attention is not large, or, in

other words, the yellow spot of consciousness is small, like that of the eye. According to Wundt, there may be four or five visual simultaneous impressions—lines, letters, or numbers. If successive, and with the most favorable interval of two or three tenths of a second, sixteen simple and eight double impressions are possible. If successive sounds be rhythmical and in groups, the largest possible number of impressions attended to at once is forty, if divided into five groups not more than three tenths of a second apart. Wundt fixes the extreme possible duration of any act of attention at from two and a half to four seconds.

3. The physiological condition of attention is a rush of blood to the nerve centers involved and the strong innervation of the end organs or of the muscles used. This is so because the centers are strained to the uttermost, and require quick, continuous and ample nutrition. Thus, in looking attentively at anything, the various ganglia in which the optic nerve is rooted are richly supplied with blood, and the end organs of vision and the eye muscles are vigorously innervated.

4. Attention is spontaneous or voluntary—you may be made aware or you may make yourself aware. It is either sensorial or reflective, directed to what is without or to what is within.

5. If spontaneous, it is caused by emotional states: we attend to this or that because for some reason we want to and are attracted. Hence, spontaneous attention reveals character; the things apperceived betray the quality and working of our emotional natures.

Surprise is such a fixation of consciousness of high intensity. We speak of a person's being rooted to the

spot, chained, fascinated, etc. As Plato tell us in his
Theætetus, " Philosophy begins in wonder "—that is,
in a spontaneous but very vigorous observance of phenomena.

6. Voluntary attention is the result of education, a
cause and an effect of civilization, a sociological phenomenon. Ribot suggests that it originated in woman, through her cruel necessity of doing unattractive
work; thus she first won the gift of application. And
Ribot is certainly correct in describing three stages
from infancy to manhood :

(1) In childhood such attention is secured by
means of education, acting only upon the simple feelings—love, fear, desire of reward, shame, etc.

(2) Later it is aroused and maintained by appeal
to feelings of secondary formation, as love of self, ambition, emulation, etc.

(3) Still later, organization comes in and voluntary attention becomes a matter of habit.

7. A measure of solitude securing freedom from
disturbance becomes the necessity of the intense
thinker or observer. Mohammed must go into the
mountains above Mecca, Paul sojourn in Arabia,
Dante haunt the woods of Fonte Avellana, and Schiller
roam by brook and glade, while Cervantes does his
best work in prison.

Voluntary attention through long habit may acquire the absorption of absent-mindedness. Archi-
medes would forget to eat his meals, and only compulsion forced him to the bath ; he lost his life in such
a fit of abstraction, at the hands of a Roman soldier
to whom he was too absorbed to return the answer
that would have saved him. Sir Isaac Newton would

sit, half dressed, on his bed for many hours of the day, when composing the Principia.

8. Intensify the attraction and so the consequent absorption, and we have the condition called rapture and ecstasy. Socrates was liable to fits of abstraction so complete that it was impossible to arouse him until attention voluntarily withdrew itself. Once in the camp at Potidæa he stood twenty-four hours in the sunshine and in the dew, motionless. The prophet Ezra sat crouching in the court of the temple from morning until night in an ecstasy of horror.

CHAPTER VIII.

THE ENCHAINING AND GROUPING FUNCTION OF CONSCIOUSNESS.

1. MENTAL states are recognized as coming and going in chains and groups. Think of Lamarck and Darwin comes into view; picture Adam and Eve promptly appears, apple in hand; hear the hum of bees and you smack your lips for honey; see a cow and you long for cream; say "one," and "two," "three," "four" come crowding on; hum a theme and an entire symphony seems to swell upon the ear; and so on *ad infinitum*.

2. There are seeming exceptions. Often mental states succeed each other without break or outside suggestion that are not apparently related. You smell an odor of jasmine and think of Mount Desert, but perceive no connection. But if you will allow your mind to dwell upon the matter, the search will

probably be rewarded by the coming up into view of certain submerged links in the chain, whose absence caused the apparent break. Then all is plain: the jasmine perfumed the handkerchief of the young lady from Boston, and the fabric of lace—borne by you on the winds from the passing yacht—you gallantly rescued from the waters of Bar Harbor.

3. The laws of this enchaining and grouping are not far to seek, if by laws we mean only a classification of the kinds of chains and groups. As these kinds are merely the conceivable relations of things—spatial, temporal and logical—we may, if we please, exercise much ingenuity in classifying. Usually philosophers have arranged them under a few captions, thus:

Contiguity . . . . Horse and rider.
Contrast . . . . Light and dark.
Resemblance . . . . Grant and Sheridan.
Succession . . . . Quoted words.
Cause and effect . . . . Vice and misery.
Whole and parts . . . . United States and New York.
Genus and species . . . . Dog and greyhound.
Sign and thing signified . . Cross and Catholic faith.

4. Sir William Hamilton has reduced these to two, simultaneity (in time and space) and affinity, the latter including every kind of logical relation. Then, following St. Augustine, he compressed these two into one, which he named redintegration, and which may be stated thus: "Those mental states suggest one another which have at some previous time formed parts of one mental state." Contiguous and successive states associate themselves because at some time joined in consciousness; and logical relations provoke associ-

ation because the mind has perceived such relations and grouped together things thus naturally in affinity. When a new fact is cognized, we note its surroundings, antecedents and consequents, and we perceive or study up its relations and then place it in its own classes; and henceforth it is likely to call out or to be called out by any member of these classes. Vice does not suggest misery until we discover that the one is a cause and the other an effect; henceforth, associated by this mental act, either may call up the other.

What, however, shall we say of the quick association of new facts with mental states—that they never could have met in consciousness? For instance, you are introduced to a Mrs. Irving Booth, and soon find yourself repeating the name of the distinguished Salvationist, Mrs. Ballington Booth, though the two have never been in thought together before. The solution is simple. The name Booth has many times formed part of the whole thought Ballington Booth, and it is that name recalls the Ballington. Henceforth Ballington and Irving, hereby associated, will be able to suggest one another without aid of the surname.

All new objects of thought must contain some quality or condition, already in some class of memorized qualities and conditions, and it is by these and their associations that what is absolutely new is joined to what is old.

5. But in the infinity of possible concurrences what is it that determines the appearance of states actually restored? Why, when I recall the song at last night's concert, do I think of the singer rather than of the programme, or of the programme rather than of the

audience, or of this or that, out of a thousand possible restorations?　It depends upon—

(1) Habit.　Familiar combinations are wrinkled into the nerve structure and tend of themselves to recur.　Frequency of recurrence in thought, for the wonted notion, establishes lines of least resistance, neural highways easily traversed.

(2) Recentness.　Poems, orations, series of facts, readily restore themselves, if but recently committed, even to the scholar of mediocre memory.　Not only frequent but recent recurrence is essential to restore them.

(3) Vividness.　This is of value because of the deep cutting in of the record on the neural tablet. Moreover, vivid mental states not only leave a more enduring record; by their very intensity they associate themselves with a larger range and variety of other mental states.

(4) Interruption.　Which may interpose sensations powerful enough not only to start new chains and form new centers of circling ripples, but also to force out of consciousness states already found in possession. Conversation is a perpetual disturbance of the associated flow of thought, a continual throwing of stones upon the already disturbed surface.　Every remark, question, or gesture of a companion starts new ripplings and establishes condensing centers for related ideas.

(5) Voluntary preference; whereby the will summons, retires, combines, disassociates and recombines the mental states.

6. The briefest association time on record (known to the author) is ·341 of a second.　A simple method

of handling this problem is to read aloud as rapidly as possible. Every word sounded is an act of association; as so many are uttered in a minute, divide sixty seconds by this number, and from the result subtract the perception time and the interval of utterance.

7. The relations of things being very numerous, the possibilities are countless. General Grant, for instance, is classed with mankind, with men, with Americans, with great generals, with Presidents, etc.; he succeeded Johnson and preceded Garfield; he was a cause and an effect; in reticence he was like William the Silent, in temperament he contrasted Washington! he was part of his army, part of his family, etc. The number of mental states which the name of Grant may revive is thus practically countless.

CHAPTER IX.

THE GENERAL QUALITY OF MENTAL STATES.

1. Mental states may be classified as initiative, habitual or instinctive; and all living matter may be said to be capable of exhibiting these three phases of mind.

2. Mind is initiative when its operations are, for the creature in question, novel; and that even the lowest forms can entertain novel psychoses is now beyond reasonable denial. It is shown in the capacity to learn, as displayed by all animals and not impossible to plants. Protoplasm, as has already been remarked, can be educated. Even bees and ants, though in popular estimate the very incarnation of routine,

have not yet passed beyond the point when their young need to be trained by their elders in knowledge of life. An ant of the slave species, if captured when a pupa, will grow up in the captor's hill in perfect ignorance of its kindred, will fight them if necessary and will learn obedience in humility. Young ants or bees all receive a certain schooling in the hill or hive. The same mental initiative appears in the shrewd devisings of various creatures to control novel circumstances, as narrated in countless trustworthy anecdotes. Commander R. H. Napier describes the feeding of a number of pigeons upon a few oats accidentally let fall by a cartman while fixing the nosebag on a horse standing at bait; all the grain at hand having been devoured, one of the birds arose, and, flapping its wings furiously, darted at the horse's eyes. The startled animal tossed his head and in so doing shook out more kernels. This proceeding was repeated whenever the pigeons had exhausted their supply. Another witness tells of two swallows who built a nest in the veranda of a house in Victoria; as the nest leaned upon a bell wire, it was frequently disturbed and twice pulled down. The pair then began afresh, making a tunnel through the lower part of the nest, around the bell wire; and they were annoyed no more.

3. This initiative, however, tends to become habitual, because of the neural law of habit; in accordance with which nerve elements can adapt themselves to peculiar functions, repeated performance develops facility and a nerve system "grows to" the modes in which it is exercised. Hence the possibility and the tenacity of personal habit. Shakespeare has observed, " How use doth breed a habit in a man!" and long before his

day Ovid wrote of evil ways and he might have said it
of the paths of peace :

> " Ill habits gather by unseen degrees,
> As brooks make rivers, rivers run to seas."

4. The early result of frequent repetition of any
act is a simplification of the necessary movements.
Habit finds the lines of least resistance, and these become trodden paths ; and hence the work is done with
increasing directness, accuracy and ease. Gradually
the act tends to become reflex or automatic, and the
conscious self is less and less troubled with care of its
supervision. When we first learn to play on a musical
instrument, to skate, to swim, to ride a bicycle or to perform some other dexterous combination of activities,
we find it necessary to regard every particular movement, and even then are clumsy and soon wearied ;
ere long, however, all these things are done without
awkwardness, fatigue or even conscious attention, the
trained nerve centers working satisfactorily under only
general supervision. Huxley tells of a practical joker,
who, seeing a discharged veteran carrying home his
dinner, suddenly called out, " Attention ! " The man
instantly brought his hands down, and lost his mutton
and potatoes in the gutter.

This law guarantees mental evolution and renders possible complex mental operations. Well says
James, " Habit is the fly-wheel of society," and we may
add, it is the condition of progress ; it forms the con-
servative factor in the growth of mind. By its help
we trail our way through the tangled forests of life's
devious experiences with ease and comparative safety ;
without it there could be no evolution of mind or
mankind.

5. The totality of habits is very nearly the sum total of personal character. Said the Duke of Wellington : " Habit a second nature ! Habit is ten times nature ! " There is an old Greek fable which declares that the goddess of Love once converted a weasel into a beautiful woman ; and it adds that this fair creature could never see a mouse without jumping at it.

Hence the personal value and the danger of habits : they represent the grooves worn into our brains by long usage ; and as they remorselessly tell the secrets of our past lives, so they peremptorily condition our future. Well said Novalis, the German philosopher, " Character is destiny." That is, our constitutional habitudes weave our fates ; they curse and they bless us.

6. Habits are inheritable. This is now denied by a large and able body of extreme Darwinians, who will allow no cause for evolution but natural selection. Over against their theory, however, there is a host of facts not easily explicable except on the old and popular belief in the heredity of habit. Take the following, and such cases are legion. Surgeon-General Hammond tells of a gentleman who, having formed the habit of taking a cup of tea at midnight, did this for twenty years. His son, born after his death, and knowing nothing of this, at twenty years of age one midnight awoke with an intense desire for tea, rose and gratified the longing; the next night the same thing recurred and it became a lifelong custom. This man died when a little son was but six years old ; the boy grew up, and seldom tasted tea, until on a certain midnight the now ancestral passion suddenly seized him and he became an habitual midnight tea-drinker. The grandson, up to the development of the

custom, had never heard of the usage of either father or grandfather. We can all recall such instances: the acquired habits of parents, whether animal or human, become inherited habits in the offspring. This is very marked in the results of the training of dumb animals. The retriever, the setter, the collie and the spaniel, among dogs, are good instances. A cross with the bulldog has affected for many generations the courage and obstinacy of greyhounds, and a cross with a greyhound has given to a whole family of shepherd dogs a tendency to hunt hares. Mr. Douglas Spalding declares that one day, after fondling a dog, he put his hand into a basket containing four blind kittens three days old. The doggy smell his hand carried set them puffing and spitting in a most comical fashion. It is evident that the antipathy to dogs was inherited, and also that in the ancestry of the kittens it had resulted not from congenital variation but from bitter experience.

7. When habits are inherited we call them instincts; and such instinct is thus an individual intelligence become racial. It is ancestral experience crystallized into race character. Le Conte calls it "communal experience treasured in inherited structure"; he defines it as "inherited memory," as "inherited knowledge." But the memory, the knowledge, the experience were become habitual and so automatic before inheritance. Long since—it may be ages ago—individual experience resulted in usage; and this usage —an ancestral heirloom—became a mental tendency. Thus there were swallows in North America before colonists arrived, and only after the land was settled did chimneys and barns become manifest conveniences for

these birds, which their individual intelligence discovered and appropriated. A habit resulted which finally, by inheritance, was crystallized into instinct, and there are now barn and chimney swallows. The same may be said of animals introduced from the old country: in manner of life and methods of chase and escape they have accommodated themselves by a play of initiative intelligence, stiffened into habit and inherited as an instinct, to their new environment. The fear of man acquired by creatures running wild in once inhabited regions is the result of surprised observation and bitter experience become a race heritage.

This whole process is beautifully illustrated in the recent life history of a small parrot in New Zealand, the kea, which until recently fed on insects and the honey of flowers. Latterly it has taken to a meat diet, and lives on sheep. It began by picking at the sheepskins hung out to dry, and at carcasses of mutton in process of curing. About 1868 it commenced to attack living sheep, which were often found with raw and bleeding backs. It has now learned to burrow into the animal's body, eating its way down into the kidneys, which form its special delicacy.

8. Instinct may work in full vigor on the moment of birth, as in the case of sucking with infants, or it may be delayed for years and then appear entirely without education in great energy, as in the instance of the tea-drinking habit just cited. A Mr. Lardner has stated, in Nature, that his brother extracted from the oviduct of a West India snake two snakelets six inches long; both, though unborn, threatened to strike, and made with their tails the characteristic burring noise. On the other hand, Spalding kept

young swallows caged until they were fledged, and then allowed them to escape; they flew off directly, showing the instinctive power of flight in a perfect but deferred form.

9. Some kinds of instinct display evidence of a high degree of original initiative intelligence. As in case of the California woodpecker, which bores holes into the bark of trees and plugs them up with wormy acorns, thus allowing the grubs within to fatten and furnishing itself with a rich future repast; or of the wasp, that stings spiders in the nerve centers, paralyzing but not killing, and so preserving them as food for its larvæ.

10. We have remarked that all inherited habits are instincts; now we must add that not all instincts are inherited habits; they may result from sexual or natural selection. Says Lloyd Morgan: " The instincts of female insects, which lead them to anticipate by blind prevision the wants of offspring they will never see—of caterpillars, which compel them to make provision for the chrysalis condition of which they can have no experience, or of the copepod crustacean, which lays its eggs in a brittle star that they may therein develop, probably in the brood-sac, and may even destroy the reproductive powers of the host for the future good of her own offspring—these and many others would seem to have no basis in individual experience." But even in these cases the instinct becomes a racial heritage; and though the impulse is too blind to be termed intelligence, as a psychic feature it belongs to the mental rather than to the vital factor.

11. The relation of instinct to initiative intelli-

gence being thus intimate, we need not be surprised
to find that the two are present in animals in an in-
verse ratio of predominance. The more instinct the
less individuality, the more inividuality the less instinct. As Le Conte argues: " The mental wealth consists of two parts—individual and inherited. In man
the individual acquisition is large and the inheritance
comparatively small. In the lower animals the individual acquisition is small and the inheritance is large.
. . . We now see why intelligence varies inversely as
instinct. It is because with high intelligence actions
are so varied in different individuals and in different
generations that it is impossible that their results
should accumulate in and become petrified in structure. But in the lower animals the conditions of life
are narrow, the habits run in few lines, and these are
deepened with every generation, until they become, as
it were, petrified in brain structure; . . . all such pet-
rifactions arrest development, because unadaptable to
new conditions."

12. Instinct may become very stupid; as is seen
in the tendency of caterpillars to go back to the beginning of a series of actions to commence over again,
when interrupted. The very wasp, which so wisely
walls up its prey in burrows, will go through the ac-
customed action of closing a burrow from which it
knows the prey to have escaped, before proceeding to
fill and seal another. The periodic migrations of the
lemming, a rodent of Norway and Sweden, has for
ages furnished amazement to the scientific world. At
varying intervals of from five to twenty years certain
cultivated districts are overrun by these little creatures; in an army they steadily and slowly advance

down from the mountains to the sea—regardless of all
obstacles, swimming across streams and lakes, devas-
tating every field, pursued and preyed upon by wolves,
bears, foxes and eagles, countless millions swarming to
the seashore; the ocean attained, they plunge boldly
into the waves and swim until exhausted they sink
beneath the surge. Doubtless in some previous age
with a different geological aspect this migration was
a movement of wisdom required by circumstances and
justified by the results: it is now only the blind working of a dangerous instinct.

Very likely the same fate would overtake European birds that annually migrate to Africa by way of
Italy and Sicily, were the African continent to disappear. This migratory habit was formed at a geological period, when there was practically a land connection between the northern and southern continents,
and when the African elephant and hippopotamus
roamed over Sicily. Should the north coast of Africa
sink beneath the waves, it is all but certain that European migratory birds would seek its sands and groves
to return no more.

What we term absent-mindedness is often only
the stupidity of mechanical thinking; as with that
Texas farmer, who drove five miles ere he discovered
that the tail-board of his wagon had been forgotten,
and returned to find, as he dismounted in his yard,
that all the while he had been sitting upon it; or
as in the case of that eminent Connecticut clergyman, who on a noted Sabbath morning forgot to
make the " long prayer," and could not understand
why the service ended at half-past eleven o'clock—a
circumstance absolutely unique in his ministry.

The stupidities of ultra-conservatism illustrate the same infatuation of habit. Truths are held to be true merely because they are not new; and institutions are valued chiefly because well established. Somebody has well said of the run of mankind, " Men are only dead men warmed over."

13. It is man's glory, however, that he may rise above the instinctive to the initiative. He is, after all, not a mere brain structure, not a nerve machine constructed and wound up years ago. He not only inherits habits; he may generate them. So doing he reigns. There are no kings and queens in the world any more save such as these. Originality comes to a throne. History of each one is always expecting formative action, and the world is to every person a constant challenge of opportunity. He who acts instinctively is human, he who lives a life of habit has formed a character, but that one who can develop new habits and bequeath new instincts to the race is divine— poet, genius, prophet; the world waits for him, persecutes him, builds his sepulchre and worships him.

14. To sum up, we find a three fold stratification of psychic phenomena:

(1) An inherited constitution of instincts, or inherited memories and aptitudes.

(2) A superadded mass of habits, or acquired memories and aptitudes.

(3) An uppermost layer of individuality, forming new memories and aptitudes.

CHAPTER X.

THE INFLUENCE OF MENTAL STATES ON ORGANIC FUNCTIONS.

1. These lectures will often emphasize the fact of influence by organic functions over mental states; we purpose to prepare for this by treating here of the reverse fact, as one of general interest. Says Prof. James, "A process set up anywhere in the centers reverberates everywhere and in some way or other affects the organism throughout, making its activities either greater or less."

Notice the influence of mental states upon the secretions. Sorrow in moderation increases, in excess checks, the flow of tears. Anxiety often occasions perspiration. The transudation of bloody sweat, in extreme mental agony, is in a few cases at least well attested as a historical fact, and entirely apart from the record of Gethsemane. The immediate and striking effect of mental states upon lactation are well understood.

Or notice the effect upon the vital functions. An instrument for measuring the rhythm and flow of the pulsation will record extreme unrest in the blood-vessels, conditioned by passing emotions; which show themselves potent in constant changes. Thus a dog's circulation exhibits tumultuous pulse-markings when listening to the sudden scream of another dog. "We catch our breath" on a sudden alarm. We "hold the breath" whenever attention and expectation are strongly engaged; and a sigh marks the relief of dis-

traction. A certain Colonel Townsend could voluntarily slow up or quicken the action of the heart.
Many persons can blush at will; and now and then one
is found who can faint if desirable. Great fright may
cause the heart to stop beating and the blood to " cur-
dle," and either joy or fear, if sudden and intense,
may occasion instant death. Excitement quickens
the circulation; modesty and shame reveal themselves
by blushing. The sight of anything horrible may induce a faint; while a disgusting object, or even the
thought of one, may bring about vomiting.

The influence of the mental states even upon the
muscles is to be noted. Maniacal fury vastly aug-
ments the bodily strength, and determination has
much to do with both vigor and endurance. The somnambulistic condition seems at times to impart astonishing acuteness and accuracy to the muscular sense
and to muscular activity. A lively play of the imagination provokes expressive movement of the features, gesticulation, and perhaps talking aloud. An
actor can only with difficulty declaim a part expressive
of intense ideas without grimace and posture. Some
guileless people record the whole inner soul in the
features and movement.

A belief that ghosts are present invariably causes
a cold shudder or the sensation of a cool draft.

2. So tremendous is this power of mind over
body, that diseases may often be cured and ailments
caused by a new idea. A woman once came to Surgeon-General Hammond with what he considered an
incurable disorder. She sighed as she turned to go
away disconsolate, saying, " Ah, if I but had some of
the water of Lourdes ! "—for she was a devout Catho-

lic. Now it so happened that a friend had brought the doctor a bottle of the genuine water of Lourdes to experiment with. He informed the patient of this, and promised her some, provided she would first try a more potent remedy, Aqua Crotonis (New York city aqueduct water). The woman consented, but protesting that this latter could not reach the case. He then gave her a little vial of the real article, but labeled "Aqua Crotonis." When this had failed he gave her Croton water, but labeled "Water of Lourdes." The result was a complete cure.

On the other hand, diseases may arise through ideas. A woman saw a child caught in a gate, and she believed for a moment that its ankle had been crushed. So deeply did sympathy cut, that one of her own ankles swelled and reddened. Dr. Morton P. Prince cites the case of a lady who believed that the mere presence of a rose in the room brought on violent catarrh and weeping; and when she smelt a rose these symptoms did invariably occur. So her physician presented her suddenly one day with an artificial rose, occasioned these disastrous results, and then confessed the fraud. The mental shock of the revelation restored her to sanity, and the affliction ceased. It was the false idea produced the symptoms; this removed, the diseased condition was gone.

3. The hygienic value of this fact is very evident, and in it lies the secret of the faith cure, mind cure and Christian Science. The Hebrews were wont to quote to one another this proverb, "A joyful heart maketh a happy cure." As persistent attention and exaggeration of ideas will account for most of our grievances and woes, so distraction from pain and

dwelling upon pleasure will guarantee contentment and peace.

4. There is danger to health and sanity in a greedy brain, which if pampered will take more than its share, in any case large, of the body's nourishment. The result is the starvation and drooping of the vital organs, and the failure of the machinery of nourishment itself. Much of ill health among students results from this overindulgence of the brain, with the inevitable final failure not only of the general physique but no less of the central ganglia themselves, whose greediness caused the trouble. It is impossible long to nourish the head at expense of the body; general decay sooner or later must set in.

5. An extremely common morbid result of undue mental anxiety is what has recently come to be called nervous dyspepsia, which is a failure of innervation of the stomach. Extremely freakish, it depends upon moods and conditions. The simplest food may fail of assimilation, and the most complex may at another time be appropriated with ease. The immediate cause is an inhibition of the nerve of the stomach, the remote cause general nervous exhaustion, or at least that irritability of brain ganglia which precipitates general exhaustion.

6. Nervous prostration—a convenient phrase covering much ignorance on the part of the physicians—in general describes the most prominent and the most alarming malady of the day. It has many forms and numberless symptoms, but its cause is exhaustion of the nerve cells, through starving or overwork. Doubtless the age is responsible. The sleepy days of former stupid discontent, when most men drowsed and the son

followed in the footsteps of the father, untroubled by
ambition, social problems or religious perplexity, are
forever gone. We are come to an age of intelligent
unrest, aspiration, inquiry and endeavor. Human action is in general intense to universal nervousness; human thought is in general without repose. The times
are feverish. There are scarce any more Sleepy Hol-
lows even among the mountains and in lonely forests;
the railroad king or the statesman is as likely to come
forth from the cot in the wilderness as from the heart
of a city. The very plow handles think. Villages are
become but suburbs to the universal civic pande-
monium. Notice how popular words which describe
popular men—"wide awake," "smart," "clever,"
"sharp"—indicate the intensity of the striving. Hence
the prevalent diseases are of the nervous order, hysteria, apoplexy, neurasthenia, brain-softening, insanity.
The phrase *nervous prostration* describes the first
monitory approaches of these insidious foes to happiness and health. It assumes protean forms, and has
numberless symptoms, the most marked of which are
incapacity for mental work, persistent depression, indigestion and insomnia.

7. The proper care of the brain involves:

(1) Its nourishment by good food well digested.

(2) The preservation of tone throughout the body
by careful prevention of an oversupply of the brain,
which should not be allowed to rob the stomach and
other vital organs.

(3) Periodic rest in sufficient daily sleep and sab-
batic and yearly vacations.

(4) All of which involves a judicious limitation of
the work done.

(5) It should be a fixed habit to divert attention from personal pain, from the foul, morbid, and horrible, and to keep the mind sweet and clean, hopeful and aspiring, stored only with the facts and fancies of the true, the beautiful, and the good.

(6) The imagination should be used to intensify the "sweetness and light" of existence. Schiller said that a truly artistic imagination "only plays with the beautiful and plays with the beautiful only." It much concerns mental health that the imagination should "play" with only the fair and winsome.

(7) Will to be well! This, strictly speaking, is the "mind cure," is potent in nerve diseases, and is not useless in other maladies. Physicians are constantly telling their patients to "give up and go to bed," but worse advice, except in the doctor's interest, could not be offered. Never give up, and do not go to bed unless to sleep.

Note.—The famous Thomas K. Beecher, in a sermon of review, stated that during a ministry of many years he had buried two thousand persons, and only three of them had died a natural death. On being reproached for so extraordinary a statement by an eminent neighboring doctor of medicine, he vented his little joke and explained by saying that those three were the ones who had not employed a physician. This will at least serve to illustrate the growing feeling among men of thought, that we have been doctored overmuch, and that the recuperative powers of the human body have not been sufficiently appealed to through the imagination and the will.

CHAPTER XI.

SUBCONSCIOUSNESS IN GENERAL.

1. By the word subconsciousness we describe mental states that are neither conscious nor unconscious.

2. We have observed that the nerve centers, directly they become "lower" and subordinate to higher co-ordinating centers, retreat into the background, their activities fading out of personal sight and surpervision. Consciousness withdraws to the higher, and the lower perform their work with an intelligence of their own that is automatic and in a measure impersonal and beyond the purview of the ordinary every-day self-recognition.

3. We have seen that even personal habits tend to retreat from the field of conscious activity and to become automatic and impersonal. The same is true of instincts, which rule the life with or without conscious supervision.

4. Sleep introduces us to another condition of the subconscious, and one in which it is possible to investigate the condition itself.

5. To this realm belong those obscure mental activities which recent writers have termed "unconscious cerebration." A man wearies of a problem he can not solve, and leaves it in despair; on the morrow, unexpectedly and when he is thinking of something else, the solution comes to him as a happy thought. You forget a name, and give up the attempt; by and by it pops into thought unceremoniously. One hears a tune, likes the air and forgets at once; on the morrow it can not be recalled, but a week later one is found humming it over. We sleep with a determination to rise at a certain hour, and on the stroke of the clock we are somehow aroused. A large proportion of our thinking and willing is done for us by a somewhat within, and we get only the results. Often the obscure decision anticipates our conscious discussion and resolution. As was true of that country parson who, called to a city church on a large salary, betook himself to prayer for light. After several weeks a neighbor accosted his eldest: "Say, Jim, is your father going to accept that call?" The boy replied, "Well, father is still praying for light, but most of the things is packed!"

6. Here, also, find place somnambulism, hypnosis and those subtile powers of the human mind which hitherto have been claimed for sorcery and spiritism, and which now we have come to name thought-transference and lucidity.

7. To these must be added certain diseased conditions which, in the decay of personality and the fading out of consciousness, push up into notice—to wit, hallucination and dual and multiple personality.

8. All these subconscious states are marked by

automatism, which is quite independent of the ego, and often defiant of it. They form a personality of their own, and develop consciousness beyond the threshold of consciousness.

CHAPTER XII.

SLEEP.

1. NERVELESS creatures do not in any true sense sleep, but they have seasons of repose that may suggest and may even simulate it.

Unicellular organisms very commonly go through stages of inactivity, when they are encysted and quiescent. Plants enjoy periods of rest, and often they droop and fold their leaves at night; and nerveless communities of animal cells are not incessantly active.

2. Animals with nerves not only can, but must, sleep. The intenser the mental activity the greater the need. Nerve cells in action consume much precious substance, dissipate enormous stores of energy and will die of exhaustion if constantly worked. The lifelong perpetual beating of the heart may seem to be in contravention of this, but who knows that this wonderful organ is innervated all through the twenty-four hours of the day by precisely the same cells? Analogy renders this extremely improbable.

3. With all animals that have active brains, sleep is a very significant factor, not only of health, but no less of life itself. In man's development it assumes vast importance. The worst form of torture for us is to be kept constantly awake. Continued insomnia

results in madness, and ultimately in death. And we
are all ready to say with Sancho Panza, " Blessings on
him who first invented sleep ! "

4. Sleep is induced by weariness, darkness, quiet
and low monotonous noises, like the buzzing of insects,
the murmur of a breath of wind among leaves, the fall
of a tiny surf upon the seashore, a mother's lullaby,
or the droning of a dull preacher ; or by gentle movements, like the rocking of a cradle and the swinging
of a hammock. In short, anything which soothes
psychic activity and determines blood from the brain
will tend to cause somnolence.

5. On the contrary, awakening can be effected by
any kind of rousement, determining blood to the
brain and exciting psychic movement. A sharp call,
a vigorous shake, or a sudden flood of light will generally suffice.

6. The process of going to sleep is very interesting. The members succumb in regular succession ;
first the head grows heavy, then the upper eyelids
droop, the subhyoid muscles yawn, the inspirations
become slower and deeper, the lower jaw falls, the
chin drops upon the chest, and the limbs relax. A
similar sequence of psychic phenomena occurs.
Speech becomes confused, vision indistinct, thought
obscure. First the will and the moral nature go to
sleep, and consciousness falls into a petty anarchy.
Visions come and go, often with marvelous rapidity,
in grotesque connection and succession, continuing
nothing, perfecting nothing and evanescent. At last
the imagination slumbers, and there is profound rest.

7. Waking reverses this process : for it is the imagination that first arouses itself to renew its chaotic

dreaming; then follows the will, reason and moral
nature; finally the eyes see and the tongue recovers
speech.

8. The physiological explanation of these facts is
simply the withdrawal or the supply of nutrition. In
sleep the brain is anæmic. The same is true of the
spinal cord; the retina also is blanched, and all the
end organs unsupplied; indeed, the nerve centers receive only a slight and sluggish flow of blood—just
enough to repair waste but not sufficient for active
work.

Hence sleep can be prevented by excitement and
by medicinal stimulants, and can be artificially occasioned by pressure on the great arteries of the neck,
or by acting through drugs upon the vasomotor centers.

9. It is probable that in sleep the mind is at all
times subconsciously active. Dreams may utterly fail
but there is a subdued self-awareness; and some nerve
cells are always on guard and practically awake. Persons in deepest repose can be aroused by a word, if
only you know what is the exciting signal. The bark
of a watch dog, the ringing of a bell, the cry of a babe
will suffice. We can sometimes appoint an awakening with ourselves and start up on the stroke of the
clock.

10. Some good work, in a quiet way, is often done
in sleep, especially if it be restless : plans are matured,
problems solved and happy thoughts evolved, as appears on the following morning. The advice so often
given concerning some troublesome aspect of life's
puzzle, " to sleep over it," is good philosophy.

11. During sleep the temperature of the body falls

from one half a degree to two degrees, the amount of
carbonic dioxide exhaled is diminished, and the amount
of heat given off falls from 112 calories to 40 (for a
man weighing 147 pounds—Helmholtz). This shows
that tissue changes are very slight.

12. The amount of sleep required is, for a child,
one half its time; for an adult, one third. Women
need more than men and among men there is vari-
ance. Napoleon could sleep and wake at will, and
needed but four or five hours : he died of exhaustion,
however, at fifty-two. Descartes required ten hours
and was incapable of efficient brain work without it.
Doubtless, in this matter of amount, both the quality
and vigor of nerve cells are involved.

CHAPTER XIII.

DREAMING.

1. SOME people never dream, or, if they do so, fail
to remember ; with most, however, at least just after
losing one's self and just before awakening, subconsciousness is more or less alert.

2. The whole nervous system, though partially in
repose, now displays a certain amount of sensitiveness.
A touch, a sound, a ray of light, a pungent odor, a
pain, a sense of heat or cold, modifies the rhythm of
respiration, determines a contraction of the vessels
of the forearm, increases the general pressure of the
blood, causes an extra inflow of blood to the brain,
and quickens the heart-beat. This sensitiveness, both
peripheral and central, combined with ample cerebral

blood supply, gives us the physical conditions of dreaming.

3. Psychologically, the supremely important fact in dreaming is the withdrawal of the personal consciousness, with its trained will and developed moral nature. Personality slumbers, the impersonal remains awake. Reality, central control and the co-ordination of ego and non-ego, all practically cease to exert influence.

4. Some claim that dreaming is the earliest and primary form of self-awareness, and that waking is a secondary state developed to meet external needs. Be that as it may, the two are radically distinct. In the former the imagination, dominated only by fortuitous association, plays at anarchy.

5. The orgy begins even before the drowsy personal consciousness is disposed of; and for a while, and indeed so long as slumber remains light, the work of fantastic creation may be controlled. The author is able in light sleep to end his dreams by an act of will if they prove unpleasant, and to continue and elaborate if agreeable. At best, however, the sway of will is weak and brief; imagination soon and easily escapes its leash, as slumber deepens.

6. Hence dreams are apt to be irrational, not regulated by the known limitations of time, space and causation. They play childishly with extension and duration, are often utterly absurd, are sometimes quite inconsequential, and not seldom vicious or darkly criminal. Miss Cobbe cites several instances of atrocious misconduct on the part of persons whose elevation of character rendered the infamy of it quite incongruous. She tells of a distinguished philanthro-

pist, an eminent jurist, who constantly committed
forgery and regretted the act only when he learned
that he was to be hanged; of a woman whose life
was devoted to the instruction of pauper children, who,
seeing one making a face at her, doubled him up in
the smallest compass and poked him through the bars
of a lion's cage; and, finally, of one of the most
benevolent of men who ran his best friend through
the body and felt extreme satisfaction on seeing the
point of his sword come out through the shoulders of
his beloved companion.

7. Yet are dreams intensely realistic, in a way.
After all is said that can be of their unreasonableness
and immorality, they are yet sufficiently actualistic to
justify the Hebrew Psalmist, when, comparing life to
a dream-troubled sleep, he said, "I shall be satisfied
when I awake with thy likeness"; sufficiently true to
life's evanescent and unsatisfactory phenomena to
point the dramatist's cynicism, when he made one of
his players declare :

> "We are such stuff as dreams are made of,
> And our little life is rounded with a sleep,
> And like the baseless fabric of a vision,
> The cloud-capped towers, the gorgeous palaces,
> The solemn temples, the great globe itself,
> Yea, all which it inherit shall dissolve,
> And, like an insubstantial pageant, faded,
> Leave not a rack behind."

This close connection of the two worlds is not to
be forgotten.

8. Dreams are determined by central or by peripheral stimulation. Of the central determination
we know little; it seems to be an automatic action of

nerve cells in the brain, sending out thought waves that cause other nerve cells to explode and other thought waves to vibrate. The locality of the starting point and the energy of the impulsion probably determine the character of the succeeding visions.

Of peripheral stimulation very little is required to decide the nature of dreams; an odor, a breath of air, the bark of a dog, a rustle, or a cramped muscle or a touch of indigestion, the pain of a wound or a disturbance of circulation, will any and all suffice to provoke elaborate trains of fantastic imaginings. A physician who applied a hot-water bottle to his feet on retiring dreamed that he was climbing Mount Etna and found the heat insufferable. Another, who applied a blister to his head, was scalped by a party of Indians. Dr. Beattie mentions a man who could be made to dream on any subject by suggestive whispering in his ear. Lobster salad just before retiring has been known to produce very lively and not always agreeable visions. The dreaming of patients in painful illness is generally distressing.

In this connection a speculation of Prof. Ladd is of interest. He claims, and seems to prove, that the dots, lines, splashes and angles which we observe in the field of vision when the eyes are closed—what the Germans name *Eigenlicht*, and Prof. Helmholtz calls "luminous chaos" and "luminous dust"—to some extent determine the form and character of dreams, and to some degree occasion them.

9. Dreams often transpire in an incredibly short space of time. A person was suddenly aroused from sleep by a few drops of water sprinkled on his face; he pictured on the instant the events of an entire life,

in which happiness and sorrow mingled, and which
finally terminated with an altercation upon the bor-
ders of an extensive lake, into which his exasperated
companion succeeded after a struggle in plunging
him.   Dr. Abercrombie relates a similar case of a gentleman who dreamed that he had enlisted as a soldier,
joined his regiment, deserted, was apprehended, carried back, tried, condemned to be shot, and led out to
execution.   He awoke as the fatal fusillade resounded
in loud report, to discover that the cause of his disturbance was a noise in the adjoining room.   Lord
Holland fell asleep when listening to somebody reading, had a long dream, and yet awoke in time to hear
the conclusion of the sentence, of which he remembered the beginning.

10. Another most interesting feature is the com-
pleteness of the hallucination.   There is absence of
all surprise over ridiculous transformations, grotesque
situations and impossible combinations.   "*Nil ad-
mirari*" is now the motto of even the most skeptical
and the most susceptible.   This is because impressions
from the outside are not present to contrast ideas:
ideas have undergone an absolute as well as a relative
increase in intensity.   Time, space, motion, pleasure,
pain, are exaggerated, and, occupying the whole field of
thought, produce upon the subconsciousness the effect
of reality.

11. The coherence of these hallucinations is worth
consideration.   While dreams do not in general
"stand upon the order of their going," there is enough
of orderliness of sequence to suggest the working of
some law of connection other than that of mere association of ideas.   Many persons, of whom the author

is one, receive their first intimation of approaching
sleep in fragmentary pictures, which succeed one
another in the most incoherent way, like views from a
stereopticon thrown upon a canvas, whereon the audience knows not what will appear next—a grand old
moss-covered castle, a tall chimney, the face of a
friend, etc. This is followed by a dramatic show with
lively action, in which the dreamer may be actor or
spectator, or both, and which, however grotesque, at
least preserves a thread of sequence. It seems highly
probable that the nerve cells of the brain, on being
loosed from control, acting at first disjointly, as slumber deepens soon begin to combine, but under some
sway less rigid than that of the conscious will. A
kind of dream personality is suggested.

12. In the same connection is the curious fact that
we dream that we dream. Often we bemoan lying
awake, when some one stirs us and we learn to our
astonishment that we have been only dreaming that
we were awake. Much insomnia is little less than this
restlessness and vivid dreaming.

We are confident that this phenomenon occurs
only with habitual dreamers, in whom the secondary
dream personality is so well developed that a fainter
tertiary personality looms up in the distant shadows.

We have no evidence that any one ever carried this
involvement into further complications, to dream that
they dreamed that they dreamed—though this is by no
means impossible.

13. Sometimes dreams manifest a vigor and range
of intelligence not usually in control. While most
persons have only silly imaginings in slumber, some
see visions that are the product of much creative

power. The author remembers on one occasion imagining himself turning over a book of fine engravings. He had never seen them before, as he assured himself while dreaming and afterward, on awaking, and while intensely vivid impressions remained. What was it enabled him in a moment to create a score of varied and superb works of art? Moreover, he has not only created whole dramas, filled with characters, scenes and witticisms; he has himself personally acted in them as one of his own *dramatis personæ* ; and then he has lain awake a long while marveling over this utterly unusual activity, never having succeeded at impersonation, nor having been knowingly capable of dramatic composition.

Some really great works of genius have arisen in this way. Tartini, a famous violinist and composer, dreamed that the devil had become his slave, and that one day he asked the Evil One whether he could play the fiddle. Satan replied that he thought he might pick up a tune, and thereupon he played an exquisite sonata. Tartini, imperfectly remembering this on awakening, noted it down, and it is now known to musicians as Il Trillo del Diavolo. And in like manner Coleridge composed his poem of Kubla Khan.

14. Dreams are occasionally significant precursors of disease. Armand de Villeneuve dreamed that a dog bit him in the leg, and a few days later fell victim to a cancerous ulcer on the very spot bitten. Gessner, in his sleep, fancied that he was fanged in the left side by a serpent; soon on the same place he developed a malignant pustule, of which he died. A man saw, in a dream, an epileptic, and shortly himself became one. A woman spoke to a person who could not reply to her

because dumb, and she awoke to find that she herself had lost the power of speech.

These facts indicate that subconscious centers are capable of sending up to the dream personality valuable information.

15. That dreams occasionally become veridical has been the belief of many in all ages. The night before Julius Cæsar was assassinated his wife Calphurnia dreamed that her husband fell bleeding across her knees. On the night that Attila died the Emperor Marcian, in slumber, saw the bow of the Hunnish conqueror broken asunder. So at least the old records tell us, and such stories are legion. Our Bible is full of similar narratives, which unbelievers have ridiculed and which the devout have swallowed with no little choking.

To the great amazement of the scientific world, the Society of Psychical Research has recently collected a very large array of no less marvelous narratives of significant dreams told by persons of the highest character and position, and verified by corroborative documents and circumstances.

We must delay the attempt to throw light upon these claims until the study of thought-transference and lucidity shall engage our attention. Well and truly wrote Byron:

> "Sleep hath its own world,
> And a wide realm of wild reality;
> And dreams in their development have breath,
> And tears and tortures and the touch of joy.
> They leave a weight upon our waking thoughts,
> They take a weight from off our waking toils;
> They do divide our being; they become
> A portion of ourselves as of our time,

And look like heralds of Eternity.
They pass like spirits of the past, they speak
Like sibyls of the future; they have power,
The tyranny of pleasure and of pain.
They make us what we were not, what they will;
And shake us with the vision that's gone by—
The dread of vanished shadows!"

CHAPTER XIV.

SOMNAMBULISM.

1. A SLEEPER dreaming sometimes acts his dream: he talks or walks. Such a person we describe as a sleep-walker or somnambule.

The condition is induced by apparently trivial causes—an undigested meal, a lingering mental excitement, or a disturbance of slumber from without. It may and often does occur during sleep by day.

2. Physiological explanation of this lies in the partial awakening of certain end organs and of the corresponding sensory-motor centers.

The psychological explanation is in the increased coherence and activity of subconsciousness.

3. Sleep-talking at first is incoherent, but it may become in time, if cultivated, a gift of intelligent conversation. One of the students of Elmira College, a remarkably talented young lady, who when awake was unusually reticent and discreet, when dreaming could be skillfully led on by her roommate to reveal all the occurrences of the day. Carpenter tells of a young lady who, when in school, often talked in sleep, her ideas always running upon the events of the previous

day. If encouraged by leading questions, she would give a coherent account of these occurrences, provided the queries were pertinent; questions not pertinent were not answered, and to all other ordinary sounds she was quite insensible.

Sleep-walking undergoes a like development. It begins in a mere locomotive restlessness, but if cultivated becomes an ambulatory life of uncanny adventure, in which certain end organs are alert and certain brain ganglia active, while the muscular system is wide awake. Sleep-walkers wander through houses, climb roofs, stray abroad over the country and in general manifest an adventurous disposition.

4. If encouraged by circumstances, the somnambulic habit develops into a secondary sleep character, a subconscious sleep life, in which the center of personality is shifted. A new memory arises; all occurrences in former attacks being tenaciously retained, a new mnemonic chain forms; each somnambulic experience connects itself with all previous ones. Moreover, the somnambule, in addition to these memories, holds in addition the entire storehouse of waking recollections, and so is richer in resources of reminiscence asleep than awake. Then characteristics assert themselves; the patient is a " visual "—that is, sees, but hears not—or is an " audile "—that is, hears, but sees not—or is a " tactile "—that is, having hypersensitive touch, dispenses with both eyes and ears. Or he may be quite like himself and only a little " queer."

5. We saw that, in dreaming, attention is not observant, but rapt; in the sleep-walking condition attention arouses itself and becomes discriminating. There is now a non-ego as well as an ego; the somnam-

bule may even perceive that he is other than himself,
and may delight in it. He may move through the
world and converse and act much as if awake, with
accurate judgment of men and things. Rational intelligence is now partially aroused, though displaying
marked departures from the normal types. Only the
old personality slumbers; the thought center seems
shifted, and a dual consciousness inaugurated. Indeed,
the somnambule often refers to his waking self as to a
third person.

6. Some very amazing incidental phenomena have
always rendered this condition the puzzle and despair
of metaphysicians and other scientists. There is often
a muscular dexterity quite unwonted, and capable of
most noteworthy feats of skill and daring, sight with
closed eyes, and touch beyond all ordinary experience
hypersensitive. Imagination is intensely vivid, and
the most astonishing creations of dreams may become
actual performances. A young, ignorant girl may begin to preach or recite poems with excellent pronunciation, rhetoric and elocution. The most intricate
problems may be solved, the most difficult music performed. We shall presently see that thought-transference and lucidity are also frequently manifested to
a remarkable degree.

7. This condition may in some patients be voluntarily induced, in which case, however, it merges into
the hypnotic trance. Of this anon.

8. Somnambules must be gently aroused, if disturbed at all. A violent shock is injurious, and may
prove fatal. In general the trance lapses of itself into
ordinary sleep, and on awaking the patient remembers the sleep acting only as a fading dream, if at all.

9. Action in sleep is much more exhaustive than mere dreaming, and the actors awake wearied and pale. Hence this condition has always been viewed by physicians as unhealthy. It would seem wise, therefore, rigorously to repress all tendencies in this direction by attention to health, the ventilation of bedrooms and the removal and prevention of disturbance.

10. A careful review of these facts will convince us that we are here dealing with only the old problem of dreaming, but in an exaggerated form and degree. Nothing absolutely new by way of psychic phenomena is developed. The creative imaginings of somnambulism are no more wonderful than the splendid visions, correct impersonations, and elevated poems and dramas of dreaming; only they are spoken and acted, as well as conceived. Its marvelous hypersensitiveness of the end organs is but what we find occasionally in the waking condition of certain exceptionally gifted persons, while its thought-transference and lucidity only multiply in number, intensity and quality, as we shall soon see—experiences which many have when in full possession of all their faculties.

Somnambulism does not offer us a new problem; it puts exclamation and interrogation points over against facts ordinarily obscure, but quite common, and vastly significant; it proves that we are creatures of marvelous capabilities; that man's knowledge of himself is the least developed of all the sciences; and that, as Sir Isaac Newton said of his own immense learning, the wisest have only "scratched the surface of things."

CHAPTER XV.

HYPNOSIS.

1. WE have learned that it is of ourselves we sleep, that sleeping we dream, and that dreaming we often talk and act our visions.

Now let sleep and dreaming, sleep talking and walking, be induced by another than ourselves, and we may describe the condition as hypnosis. We are hypnotized, the sleep is enforced and the dreams are suggested by another. In short, hypnosis is induced sleep, induced dreaming and induced somnambulism. The essential feature is the induction, and the important problem of hypnotism is the secret of its nature and method. We may expect all hypnotic phenomena to group themselves under the three heads of sleep, of dreaming and of somnambulism.

2. Hypnosis is a widespread possibility. Its range is as extensive, probably, as the possession of brains, or of elaborate nervous systems; and it thus appears that even in low forms there exists a realm of subconsciousness. Hypnotic results have been obtained in the shrimp, crab, lobster and sepia; in the cod, brill and torpedo fish; in the tadpole, frog, lizard, crocodile, serpent and tortoise; in some birds; in the Guinea pig and in the rabbit. It is generally found sufficient to place the animal in some abnormal position—for instance, on its back—and to keep it quiet, with slight, continuous pressure. Soon it refrains from voluntary movement, and anæsthesia of skin and mucous membrane results. With repeated ex-

periment animals become more and more susceptible.

3. Of human beings the majority may become
either agent or sensitive. Authorities vary as to the
proportion of sensitives ; they agree, however, in holding that persistent experiment would overcome resistance in most cases. Forel thinks that every person not insane in time would succumb ; and as in
every one there is a realm of the subconscious, this is
probably true.

It has generally been supposed that health, culture and intelligence, affording self-control, favored
resistance ; and that a weak will, incapacity to fix attention and the hysteric temperament predisposed to
easy surrender. Moll, however, with many other able
experimenters, now claims that the weak and hysterical are, if anything, less amenable to suggestion, and
that the best sensitives are vigorous in mind and body.

4. The condition is produced by any method that
fixes the attention and arouses expectation of its occurrence. Thus, by passes or other manipulation—
by causing the sensitive to gaze fixedly at a bright
object—by a sudden flash of light, a violent noise, a
word of command, etc., the suggestion may enter by
any of the senses. In well-trained cases a simple
direction by letter, by telegraph, or by telephone will
do ; and a mental command, working by thought-
transference over miles of distance, has been known
repeatedly to succeed.

Bernheim's method is as follows : " You place the
patient in an armchair, and make him for a few
seconds, or minutes, look up into your eyes ; and
meanwhile tell him, in a loud and confident but monot-

onous tone, that he is going on famously, that his eyes
are already swimming, that the lids are heavy, and
that he feels a pleasant warmth in legs and arms.
Then you make him look at the thumb and first finger of your left hand, which you gradually lower, so
that the eyelids may follow. If the eyes now close of
themselves, the game is won. If not, you say, 'Shut
your eyes,' and proceed with suggestions."

A lady accidentally hypnotized a girl—a perfect
stranger—whom she met in a railroad station, and
whose face she simply stroked in sympathy. A gentleman hypnotized his babe by playfully shaking his
finger at it. Esdaile succeeded with a blind colored
man, by gazing silently upon him over the wall, as the
patient was engaged eating his dinner; the laborer
gradually ceased to eat, and in a quarter of an hour
was perfectly entranced and cataleptic. This was repeated at untimely seasons, and when the operator's
presence could not have been known, always with like
results. The "evil eye" of ancient superstition in
this experiment was probably realized.

Baron von Shrenk-Notzing has shown that hypnosis may be hastened or intensified by narcotics. A few
whiffs of chloroform will put even an obstinate patient
into susceptible condition, and often a narcotic of itself is sufficient to predispose to all the well-known
conditions; moreover, a person only slightly under
control may be thrown into deeper mesmeric trance.
This is so because narcotics affect the conscious personality, but leave the subconscious largely, if not
wholly, awake.

5. Hypnosis occurs in varying degrees of complete-
ness, in some cases resembling ordinary drowsiness,

in others effecting profound revolution in the workings of the nervous system. A few sensitives retain
throughout somewhat of personal consciousness, and
decided power of resistance to absurd, disagreeable, or
immoral suggestions. Most, however, are only subconscious and passively obedient.

In its most perfect manifestation the condition
presents three phases, though not in any fixed order—
lethargy, catalepsy and somnambulism.

6. Lethargy, or deep hypnotic sleep—for it is
nothing else—can be produced by firmly closing the
eyes, if the entranced patient knows that such result
is expected thereby. Sensitives become under this
treatment perfectly inert, with tendency to rigidity.
Psychosis is intensely subconscious, and some claim
that the state is one of complete unconsciousness.
They, however, hear, understand and respond to the
commands of the operator, and are therefore so far
forth subconscious, precisely as in the case of ordinary slumber.

Pressure on tendons will render associated muscles
inflexible; the whole body can be stiffened by pressure on certain parts of the legs; so that patients can
be placed with head on the back of a chair and feet
on the floor, unyielding as a board. There is complete
insensibility to pain, and needles may be inserted even
into the quick between nail and finger tip without
provoking outcry. Latterly, the lethargic stage has
been used as a substitute for anæsthetics in surgical
operations; and the most elaborate cuttings have been
carried on with an insensibility as complete as that
conferred by chloroform. Sensitives must, however,
be trained by repeated mesmerizing for this.

7. Catalepsy, or hypnotic dreaming, is produced by simply opening the eyes of the lethargic patient. There is now a sort of impersonal consciousness, which replaces the coma of lethargy. An attitude or a movement can be impressed from without upon the subject, who will retain the attitude or complete the movement. In fact, the dreaming condition now obtains, only the dreams are suggested and guided by the will of another, and the patient is sensitive to only one suggestion at a time—a perfect automaton. She is a devout Catholic—for we describe an actual case—and a gong rung to simulate the ringing of a church bell will produce an attitude of prayer, with eyes lowered, and head and body meekly bowed. In-sert a red glass between her staring eyes and the light, and she will receive a suggestion of conflagration, will see flames and burning wretches, and will wring her hands in horror, fear and pity. Whistle a waltz, and she will dance. Indeed, any vivid idea suggested works its way and by channel of any of the senses, into her brain, and arouses, by purely reflex action, pos-tures, gestures and cries appropriate.

It is a curious fact that the two halves of the brain can be operated upon separately, by the direc-tion of suggestion to the right or to the left end or-gans. Thus a double and opposing suggestiveness may work the most contrary emotions and expressions; the right brain may be frightened and the left en-couraged, with ·the result of the left half of the face exhibiting terror and the right wreathing itself in smiles.

8. In hypnotic somnambulism the sensitive is sleep talking and sleep walking and knows her own

dreams. There is not only subconscious activity of a high degree of acuteness, but also a pronounced sleep personality. Sensibility to pain is fully recovered, complex ideas possible, speech regained.

Three series of phenomena are now to be observed :

(1) The sensitive is curiously *en rapport* with the operator; her sleep personality is in strange identity with his own. There is a blind confidence, a devoted clinging, an implicit trust, entirely non existent before; and so close is this psychic union that salt or pepper on the operator's tongue will cause the patient's face to draw awry, and headaches and toothaches can at a word be transferred backward and forward. Thought-transference becomes an exceedingly easy channel of communication, and commands may be made and executed by silent volition.

(2) An expressed judgment of the operator is accepted as fact; nay, more, the sensitive's imagination plays with and elaborates the most improbable assertions with infantile credulity, and as though her mind were only an annex of his own. A file bitten is pronounced good chocolate, because so declared. The patient is asked whether she hears the canary sing, and enlarges upon the variety and richness of the tones. She is assured that an Englishman present is a Chinaman, believes it, and pictures vividly his Oriental robes, slit eyes and pigtail. Another guest is accepted as a block of ice, with flowers growing on the surface; and she points to the glacial streams flowing from him, and picks Maréchal Niel roses from his pencil case. She is told to sleep, and is in profound slumber; she is awakened, and then again bidden to

7

sleep until the hat of one of the company be removed, and, obeying, the moment the hat is removed she awakens herself. She is commanded to poison the Chinaman with arsenic, does so, and weeps bitterly in remorse; giving him the phantom cup, she gasps, "Drink it not—the cup is poisoned!" as if driven by dread fatality to what horrified her.

(3) Finally, the operator may play upon the sensitive's machinery of inhibition and acceleration. He may make it impossible for her to pronounce the letter A, and take from her the very idea of the letter, so that all words containing A are sounded without it. He may inhibit the use of any sense, rendering her blind with open eyes, deaf, dumb, without taste, or without smell. He can make her lame in arm or leg. Or, on the contrary, he may accelerate any sense or function. She will detect a particular quarter of a dollar from twenty such, simply by weight, poising them upon a finger. She can be brought to see things microscopically small or through a cardboard, and even behold her own image on a piece of writing paper, using it as a mirror. Inform her that a picture is on a blank sheet of paper, and she will at once perceive it, and if it be mixed with other blank sheets will extract the right one; turn it upside down and she will complain that her picture is reversed. Nay, these imaginary sketches appear to respond to all the laws of optics, can be rendered double by pushing inward one of her eyes, can be doubly refracted, etc. Draw a mark with red chalk on paper, and assure her that the page is blank, and she will not perceive it. To be sure, the eyes work normally, and the usual impression is made upon the retina, as appears from the fact that when

her vision is directed to another blank sheet she does see the complementary green after-image of the red chalk mark. She sees and is inhibited only from perceiving. If the mark be doubled by a prism, she will perceive the second image, but still ignore the first and direct one.

9. Hypnosis is terminated by reversing passes, if these were made at first, by blowing upon the eyes, by a word of command, by predicting that at such a moment or on such an occurrence the patient will awake. Care should be taken to remove all unpleasant previous suggestions, to tell the sleeper that the situation is a very comfortable one, and that the awakening shall be to health and peace of mind.

10. The condition is favorable to the display of thought-transference and lucidity; of which a little later.

CHAPTER XVI.

THE HYPNOTIC SLEEP PERSONALITY.

1. HYPNOSIS may be repeated indefinitely; and with repetition the sensitive becomes more and more susceptible, as to speediness of subjugation, as to intensity and as to duration.

In some cases the control can be maintained for long periods—for months, and even for years.

2. Renewals all connect themselves with previous experience in a consecutive order, and a mnemonic chain forms quite as in natural somnambulism. Recollection of what has taken place in the trance rarely presents itself on awaking, except perhaps as a fading

dream; but it is perfectly active and accurate when hypnosis is renewed. Hence facts acquired in the sleep may be recovered on the awakening by indirect methods, appealing to this coherent subconsciousness. Give the aroused patient a planchette, and the needed information will be forthcoming. Mr. Gurney describes a large number of experiments in arithmetical problems given a patient when under influence, the answers having been duly written out by planchette in the normal condition, when the latter was wholly unaware of what he was doing. Dr. Proust describes a person who falls asleep himself without outside suggestion and without warning, who for short periods exists in an entirely anomalous life; he is a veritable Dr. Jekyll, only his Mr. Hyde is not at all a demon. On May 11, 1889, he was breakfasting at a restaurant in Paris, and two days later found himself at Troyes. Of what had happened during the interval he could remember nothing; he recalled, however, that before losing his primary consciousness he had worn a greatcoat containing in a pocket two hundred and twenty-six francs. He was hypnotized, and at once gave a lucid account of his somnambulation, of his visit to Troyes, of friends dined with there, and where he left the overcoat and purse. These statements were all verified, and the coat and purse with exact amount of money recovered. Other similar authentic cases are on record.

3. These facts account for the gradual rise in hypnotics of what has been called secondary personality, which, after all, is only intensified sleep personality. Chronic cases slowly develop a distinct subconscious character, and if the waking character be weak or vicious, the "new creature" may become the most re-

spectable member of the firm. (Read account of Blanch Witt and of Marceline R—— in Proceedings of the Society of Psychical Research, vol. xv, pp. 216, 217, 219, 220.)

4. When the secondary personality has become well established, itself may be hypnotized, and so in time a third sleep character appear, the shade of a shadow, the dream of a dream. This is verified in the famous case of Madame B——, an elderly French peasant, who, though old, dull and ignorant, shy, passive and stolid, has become the most interesting woman in Europe.

Leonie B—— (long the favorite sensitive of Prof. Janet), who falls asleep at a word or by volition exerted over great distance, has developed, when in the mesmeric trance, an extensive mnemonic chain and a distinct character peculiar to the condition. When hypnotized she calls herself Leontine, and is, as such, vivacious, saucy and not very truthful; her memory now is more extensive than Leonie's, comprising as it does all that the latter knows and all that the former has experienced. One day Janet received a note from Leonie written in serious and respectful style and declaring that she was ill. Over the page began another epistle in a quite different style. "My dear good Sir: I must tell you that B—— really, really makes me suffer very much; she can not sleep, she spits blood, she hurts me! I am going to demolish her; she bores me; I am ill also. This is from your devoted Leontine." Madame B—— knew nothing of this second letter when closely questioned. These duplex letters became common. Madame B—— would write Leontine's postscripts automatically, in a fit of abstraction, and if on arousing herself she discovered what she had done, she would tear up the missive. Hence Leontine hit upon a plan of placing them in a photographic album, into which Leonie could not look without falling into catalepsy.

After Leontine's personality was well established, noticing that there was a background of subconscious cerebration even in her psychic life, Janet succeeded in throwing her into mesmeric trance and thereby causing to emerge a third personality, at first faint but now daily becoming more and more characteristic. This third Madame B—— called herself Leonore, and knew, in addition to her own memories, all that Leonie or Leontine recollected. Leonore is thoughtful, grave, addicted to poetry and much the most estimable member of Madame B——'s copartnership.

5. It is said that back of Leonore a still other individuality looms up; and the question arises, whether there is any limit to these mnemonic chains and more or less distinct personalities. To meet such astounding facts, F. W. H. Myers has broached a hypothesis, in which he assumes that every cell in our bodies has its own cellular personality with its own particular memory, and that every combination of cells in or associated with limbs or organs develop composite personalities with associate memories. "These, however, do not deserve the title of separate personalities (except in the sense in which that word may be applied to the brute creation), and their memories may never come into the human consciousness at all. Above these rises the immense nervous apparatus, which corresponds to the human mind; and of this apparatus we habitually use only such proportion as our English

vocabulary bears to all possible combinations of the alphabet. The letters of our inward alphabet will shape themselves into many other dialects; and many other personalities, as distinct as those we assume to be ourselves, can be made out of our mental material. . . . Each of the personalities within us is itself the summation of many narrower and inferior memories. It is conceivable that there may be for each man a yet more comprehensive personality, which correlates and comprises all known and unknown phases of his being."

It would be premature to accept this hypothesis as anything more than a mere surmise, but its conclusion seems in the highest degree probable. That man, in the last analysis, is an indefinite series of personalities, is as utterly repugnant to self-respect as it is inherently improbable. We are safe in concluding that the facts of sleep personality emphasize not the mere divisibility of man, but the boundless resources of his mind and the countless possibilities of his being for achievement and for character.

And the supreme psychological importance of hypnotism lies in the fact that it furnishes a method for cleaving the strata of consciousness, for analyzing the workings of the mental machinery, and for studying in detail the mental processes. Like the microscope in histology, the telescope in astronomy, or the spectroscope in spectrum analysis, it is a new instrument of research.

6. While memory of what occurs in the mesmeric trance generally fails to persist into the waking state, commands and suggestions for future action, then received, are likely to be executed in due time—not as the urgings of another will but as self-suggested. A

clerk, having been hypnotized, was told that two and two make five. The next day all his accounting went wrong, and it was discovered that wherever two and two came together he added them as five. Here the sleep consciousness overruled the normal working of his mind. Promises made in the trance are performed on awaking, not as promises but as irresistible impulses. Commands are obeyed under ill-defined sense of obligation, while suggestions become happy thoughts that demand heeding.

Moll states that the longest time of successful post-hypnotic suggestion is recorded by Liébeault—one year.

Where the suggestion is not carried out duly, the idea of it remains to torment the victim.

7. Hence the therapeutic value of hypnotism. Forel tells of an old drunkard and would-be suicide, whom he recovered: in the trance it was suggested to him that ardent spirits were a curse, and he was commanded to abstain. In consequence, when awake, he, seemingly of his own motion, became a total abstainer. This experiment has proved successful in many similar cases. Habitual aches, mostly neuralgias of various kinds, have been allayed in numberless instances. Baierlacher claims to have removed pain even in a case of cancer of the stomach and for days· Chorea and hysteria also seem amenable to this treatment. Bickford-Smith transferred a headache from Leonie B—— to himself, simply by so willing.

These facts suggest an obvious danger. The subject is slave of the operator, and may become his helpless victim or ready accomplice in vice or crime. It is evident that the practice should become matter for the

strictest legal regulation. Liébeault advises, as a preventive of undue or unrighteous influence, that the
person enslaved should seek rehypnotization at the
hand of a thoroughly trustworthy operator, who is to
suggest that no other party shall have power to induce
the condition. This, it is claimed, will work complete
deliverance from the bondage.

8. Two incidental dangers beset this method of
experimentation :

(1) Cross-mesmerism—where the patient is brought
under the influence of more than one person at a single
sitting ; when with sensitive subjects there are violent
contortions and refusal of obedience to suggestion and
command. From this state it is difficult to arouse the
patient, and headache and physical discomfort result.

(2) Imperfect awakening or oversudden rousement, in the first place subjecting the patient to all
the discomforts and mischances which may befall a
person not in full possession of normal consciousness ;
and in the second, startling and shocking the nervous
system.

9. The general effect upon health is in dispute.
Probably it injures some and benefits others. The
author would advise that it be not resorted to without
cause, and that all aimless and frivolous experimentation be strictly prohibited.

10. Hypnosis only develops, in the fullest degree,
the natural possibilities of the subconscious. Its
lethargy is but deep sleep under control ; its catalepsy nothing but intense dreaming, when the visions are
suggested by another imagination ; its somnambulism
ordinary sleep talking and walking, directed by another will. The powerful sway of the operator finds

ample analogy in what the world has ever recognized
as in the scope of personal influence. Hypnotic hallucination is only exaggeration of a perfectly normal
process which tends to go on in all of us, and is repressed only by memory, and a will trained by experience. Nor are its grander performances entirely without parallel ; its outbursts of genius have been equaled
by similar extemporization in dreaming, and by the
accomplishments of the waking state, in exceptional
persons. The fact merely indicates that very remarkable developments in multiple consciousness have long
since been studied under the phrase " unconscious cerebration." Socrates had his " dæmon," and many
men have exhibited two contrasted natures. There
was once a Swedish king who entered a ballroom, in
the glow of healthful youth, to receive at the entrance
a note warning him that his life was in danger. He
tossed the missive contemptuously aside, only to fall a
little later, and in the height of the festivities, under
the treacherous blow of the very friend who penned
the warning. The wretch in one breath would slay
him and would deliver him, at once best guardian and
cruelest foe. Lacenaire, a famous French criminal,
the same day he committed a murder risked his life
to save that of a cat ! It is said of Robespierre, that
even while he was the terror of France, and wading
knee-deep in the blood of innocence, to the two sisters
with whom he boarded he was a modest, virtuous and
estimable gentleman, and they mourned his loss sin-
cerely. Indeed, we find hints of the same fact in
quite rudimentary forms of automatism. James states
that, in a perfectly healthy young man, who can write
with the planchette, " I lately found the hand to be en-

tirely anæsthetic during the writing act; I could prick
it severely without the subject knowing the fact.   The
writing of the planchette, however, accused me in
strong terms of hurting the hand.   Pricks on the
other nonwriting hand, meanwhile, which awakened
strong protest from the young man's vocal organs,
were denied to exist by the self which made the
planchette go." Binet has shown that in every one,
at all times, subconscious potentialities exist, and can
be aroused, interrogated and educated.

In short, hypnotism offers no new field of research,
but only a new method of exploiting facts which,
without it, must be at least suspected.   Some one has
said that there is nothing new in hypnotism but the
name.

11. The ethical bearings of the subject seem at
first confusing; but remember that somnambulism is
only dreaming exaggerated, and that hypnotism is
only somnambulism exaggerated, and the darkness
will clear.   One's accountability for wrong-doing in
dreams is manifestly limited to the sinfulness of previous errors in diet, and to the general trend of character; and in somnambulism it is evident that neither
merit nor ill desert can attain a high degree.   The
same must be true of the hypnotic condition : the sensitive is so completely under controlling influence as
to be practically *non compos*, and can be blamed only
for such utterances of character as without compulsion
flow forth from the nature within.   In the stage of
lethargy the patient is mere wax in the hands of the
molder; in catalepsy, the dream that goes on is controlled by another; and even in the somnambulistic
phase, so powerful is the suggestion of the operator

that resistance is generally useless, and hence responsibility at the lowest point imaginable; and courts have
so decided again and again. In cases of multiple personality, freedom of action under the new conditions
may develop, in time, into a self-control so genuine,
and a life so varied, that accountability begins to re-
cover its normal degree; it is much as in the case of
one who should be, during different periods of his life,
a clergyman, a horse-jockey and a submarine diver;
though the same standards can not be used for each
and all of his life phases, a just judgment of his substantial personality can conceivably be formed. Leonie,
Leontine and Leonore may exhibit different facets of
character, but the three are substantially and ethically
one and neither is soulless.

12. This subject should not be dismissed without
some reference to its very interesting history.

Hypnotism formed the ancient stronghold of necromancy and sorcery and in all ages has been the instrument of priestcraft, charlatanry and superstition.
Among old-time peoples, as now amid barbarous and
savage races, the ignorant and credulous were hypnotized, frightened, swayed, cajoled, injured and cured,
by frauds and illusions innumerable. Sorcerers were
both deceivers and themselves deceived : they dealt
in the dreams, thought-transference and lucidity of
hypnosis; they naturally were feared, courted and per-
secuted. Their necromancy, so far as it had any substratum of fact, was based on what has been described.

The "evil eye" was nothing but the mesmeric
glance of the sorcerer, marring by command and suggestion the life of the hypnotized victim.

Crystal vision was a picturesque form of the working of the same phenomena. A cup of ink, a crystal polished, a mirror, or even the thumb nail, was used; any reflecting surface would do, but crystal was preferred. A boy or girl thrown into sleep was set to gaze upon this surface, and on it dreams appeared, rendered objective by hallucination. Sensitives could often mesmerize themselves by simply gazing into the crystal, with resulting dreams, apparently externalized.

The sibyls and other oracles of antiquity were only sensitives who had displayed unusual gifts in thought-transference; they were always hypnotized, unless able of themselves to fall into the trance. Sorcerers secured lads, often by violence, put them to sleep, and forced them to see and prophesy. The girl of Philippi whom Paul freed was the wretched victim of such scoundrels, of whom Elymas and Simon Magus were fair specimens. Oracular performances occurred amid much impressive incantation—a darkened room, lamps burning low, clouds of incense, the sorcerer in flowing robes of much splendor, and the like.

The black art of the Middle Age was only a restoration of ancient practices under Christian auspices, with a new vocabulary. Nothing occurred that science is not to-day studying under conditions favorable to solution.

The witches of those times and later were but evil women, who, finding that they possessed a power they themselves deemed Satanic, used it to annoy their neighbors, to vent their spites, to earn a dishonest living and to make themselves feared. It was wrong to hang them, but most of them richly deserved to be hanged.

The history of sorcery, from the beginning until now, has been a dreary record of gross immorality and cruel wrong, the strong ever preying on the weak. The evils involved have always seemed so many and so great, that in all ages and lands it has fallen under the ban, in general forbidden on pain of death. Still, so prevalent has superstition proved, and such powerful interests have antagonized repression, that prohibitory statutes have been always and everywhere more or less evaded.

The first *quasi*-scientific attempts to investigate, describe and classify the facts were made by Mesmer, in association with much magician's trumpery. His crude work proved of little value except to goad scientists to accurate study of the phenomena.

In 1845 Baron von Reichenbach announced the discovery of a new imponderable force, which he called *odyl*, and supposed to exist throughout the universe, and to be developed by magnets, by certain crystals, and by human bodies. Persons sensitive to *odyl* saw luminous phenomena near the poles of the magnets and about the bodies of others in whom the force was concentrated. Hence the term *animal magnetism*, which attached itself to the whole class of hypnotic phenomena.

Two American lecturers in 1850, broaching a new theory based upon electrical discoveries, substituted the title of *electro-biology*, which became popular.

Braid, a surgeon of Manchester, first subjected the facts to accurate study in 1842, and the science of hypnotism received from him both its name and its respectability. Carpenter became his faithful expositor, examining and verifying his experiments and results. To-day many men of great shrewdness and some of eminence are pondering with deep interest the facts.

CHAPTER XVII.

THOUGHT-TRANSFERENCE.

1. THIS remarkable psychic fact has long been anticipated in discovery by certain proverbs based on an obscure perception of the general law ; as, " Think of an angel and you shall see his wings," or " Think of the devil and you shall see his horns." Also by certain facts never well understood, as the power most persons possess of disturbing another's flow of thought by steady gaze, though the head of the observed be quite averted, or the instantaneous occurrence of identical ideas or words to two or more persons when together. The discovery itself, however, has been recent, and more so the demonstration.

Thought-transference is now accepted as one of the subconscious gifts of the human mind, very generally by those scientists who devote themselves to consideration of psychical phenomena. It is freely used as a good working theory in explanation of yet more occult facts.

This theory is, that, subject to certain laws mostly unknown, thought leaps from one mind to another mind by processes unexplained.

2. The phenomena occur most persistently and vividly in the hypnotic trance, when between operator and sensitive the freest mental interchange takes

place : the former may simply will the latter to sleep
or to wake, to do or to forbear, without sign ; and even
salt or pepper on the operator's tongue will cause the
patient to make a wry face. One instance, of thousands at hand, is that of a hysterical girl of fourteen,
whom a certain Dr. Dusart could will into a mesmeric
trance and arouse without a word of command. More
than a hundred times he did this with perfect success.
On one occasion he left her without giving the usual
order to sleep until a particular hour next morning.
Remembering the omission, he issued the order mentally when at a distance of seven metres from the
house. In the morning, when he asked the patient
how it was that she had slept without any command,
she replied, "True, but five minutes afterward I
clearly heard you tell me to sleep until eight o'clock."
He then told the patient to sleep until she received
command to awake, and directed her parents to mark
the exact hour of the awakening. At 2 p. m. he gave
the order mentally, at a distance of seven kilometres,
and afterward found that it had been punctually
obeyed ; and this experiment was successfully repeated
at different hours.

This explains, in part at least, the power of spirit-
ualistic mediums. Falling into trance, or at least a
similar subconscious condition, they read the minds
of sitters, and this forms their chief stock in trade.
The author once, when four thousand miles from
home, sat with a medium, an ignorant sailor and
a total stranger, who gave him correctly his own
name and the names of mother, wife and wife's
father, besides fishing up many facts and names quite
forgotten.

3. But hypnosis is not a necessary condition; the phenomena may characterize any state or semistate of subconsciousness — dreaming, reverie, or unconscious cerebration. This will best appear in a tolerably full account of the most complete and carefully guarded of the many demonstrations. Rev. P. H. Newnham, vicar of Maker, Devonport, a respectable and intelligent English clergyman, in 1871 instituted a series of experiments covering a period of eight months. The results appear as forty manuscript pages of notes, and among these three hundred and nine replies automatically written by Mrs. Newnham with planchette, to questions which she did not see nor hear and of which she could learn only telepathically. The wife always sat at a small table in a low chair, with eyes shut, leaning backward. The vicar sat about eight feet distant, at a rather high table, with his back to the lady. Planchette would begin to write instantly, the answer often having been half finished before the question was completely written. Often the sensitive would touch the board with but a single finger, and this would suffice. She had no faintest conscious knowledge of what was in process of writing, and often no hint as to the subject or drift of the questions. The answers were a curious combination of knowledge and ignorance, never beyond the mental powers of the percipient, but on a lower moral level than her usual conversation. She would evade and even lie, when unable to respond correctly; though evasion and falsehood were utterly foreign to her character. The answers often did not correspond to the opinions or expectations of either party. Here was instantaneous and accurate thought-transference extending over many

8

months, through more than three hundred formal trials, all successful.

4. On the range of these phenomena we can not speak positively, so little is known of the conditions necessary or favoring. Some persons seem more susceptible than others. Twins are thus *en rapport* to a wonderful extent. Probably all have the gift as a latent potentiality; with a few it is a very luminous feature.

5. No limit has ever been fixed to the distance over which transference can be effected : in the case of twins it has covered a separation of thousands of miles. Nor has any light yet been shed upon the nature of the medium nor upon the speed of transmission.

6. The philosophical bearings of these facts are wide-reaching and important. As thought-transference can not be classed with sensations, and indicates a quite other inlet for human knowledge, the sensationalism of Hobbes, Locke and Comte seems annihilated. The tendency of philosophy in both France and England for over three hundred years has been sensationalistic; and nearly all the metaphysicians of these two countries have built their systems upon the postulate, that human knowledge comes only through the senses. It now appears that we must push out the stakes and lengthen the cords of our canopy of thought. It would seem that knowledge may enter unawares, in accordance with laws of which sensationalists and positivists have known nothing.

This discovery removes from the theological doctrine of a divine inspiration the stigma of violating probabilities. Inspiration has become the most fea-

sible and natural of religious processes; indeed, it is
no longer even an unlikely phenomenon. That an
Intelligence above us should drop thoughts into the
human mind seems the simplest and most reasonable
method of communication between the seen and the
unseen worlds.

CHAPTER XVIII.

LUCIDITY.

1. CORRELATED with thought-transference we
have the very different though no less amazing fact
of lucidity, or "second sight," which seems to be the
working of a supersensuous subconscious vision, discerning matters utterly beyond the reach of any known
organ.

2. History gives us instances of the exercise of
this kind of knowledge, but until recently science has
treated them with contempt. Gregory of Tours tells
us that Ambrose, having fallen asleep while saying
mass in the cathedral of Milan, dreamed that St. Mar-
tin had just died at Tours, in accord with the exact
facts. Swedenborg claimed to have seen the great
fire in London while it progressed, and though in
Stockholm at the time.

3. This gift also, like thought-transference, mani-
fests itself in a marked degree during the mesmeric
trance, and is an important feature of genuine medi-
umship. It has given to heathen sorceries and the
art magic, to soothsaying and crystal vision their uncanny significance; and it has always been a question

whether dreams do not derive a grave meaning at times from its exercise.

4. The evidence has of late years accumulated, and that such a gift is possessed by some at least seems no longer questionable. Madame B—— (Leonie-Leontine-Leonore), when hypnotized, possesses this power to a remarkable degree. Janet, Richet and other experts have subjected this woman to every variety of test. We give in Richet's own words his methods of procedure and calculation of chances: "From the midst of ten packs of fifty-two cards each, I drew at hazard a card, which I placed in an opaque envelope. I did this in low light at one end of my library, which is nearly five metres in length, Leonie sitting at the opposite end, with her back to me. . . . The envelope was gummed, and I closed it at once. . . . The name of the card indicated by Leonie was written by her in full, or written by me before the envelope was opened, and I kept an exact account of all the experiments made. No conscious or unconscious, mental or nonmental suggestion could be made by me, since I was totally ignorant of the card placed in the envelope."

Thus proceeding, in sixty-eight trials Leonie seventeen times offered full description. Of cards entirely right, with an antecedent probability of only one or two, Leonie guessed twelve; of cards with suit right, with antecedent probability of seventeen, she guessed forty-five; of cards with color right, with antecedent probability of thirty-four, she guessed forty-five. The chances in favor of this, not allowing for a law of lucidity, were one in one billion of billions. Bickford Smith, a wealthy English gentleman, who was permitted to hypnotize Leonie, asked her for a description of his father's country house in England. This she gave, with minute particulars in every regard correct. She expressed surprise at the size of the kitchen and the number of the books in the library. She placed several peculiar trees, and described the gardeners and other underlings at their work.

On the 27th of a certain September, Leonie described a bicycle race, which did not occur until the 29th ; she named the winner, and said that there would be three prizes for him—a fact no one could reasonably have anticipated ; her improbable prophecy was fulfilled by a telegraphically added prize from the minister of war, which, with what was called the " lap prize," made three in all. Cases as remarkable multiply.

Braid, of Manchester, an unimpeachable witness, narrates the story of an ignorant girl unacquainted with music and the grammar of her own language, who, hypnotized, in his presence sang songs in foreign languages with Jenny Lind, with a pronunciation and intonation so exact that persons not very near supposed there was but one voice, and that the Swedish Nightingale's.

5. Lucidity, also, aims a deadly blow at the sensational philosophy. If a clairvoyant may learn the markings of concealed cards, see visions of what is yet to occur, describe houses and people hundreds of miles away, " speak with tongues," and anticipate the refinements of finished art, then the popular schools of modern psychologists are without a philosophy. Moreover, these facts, like those of telepathy, serve to remove a reproach of long standing from the Hebrew Scriptures, which have been arraigned by scientists

for recording unnatural displays of psychic power on the part of the prophets of Judaism. The prophetic insight of the Hebrew seers can no longer be stigmatized as unnatural. They surely saw visions and dreamed dreams; the distant and the future appeared to them as a shifting panorama; and if their beholdings proved viridical, the facts did not contravene what we now know to be the bounds of reason.

6. Lucidity and thought-transference will account, in part, at least, for the rise of religion among primeval savages. The seers became prophets of mystery, and in time rose to some little glimpse of the moral order of the universe: at first mere medicine men, deceiving and deceived, they slowly ascended into a lordlier realm of spiritual insight and religious guidance.

CHAPTER XIX.

HALLUCINATION.

1. WE must distinguish hallucination from illusion. One may be deceived by his diseased senses—as when, during the sufferance of a cold in his head, he smells smoke constantly, and goes through his dwelling in search for fire; as when, if a victim of catarrh of the middle ear, he hears drums, gongs and bells sounding loudly, and now and then is startled by the distinct calling of his name; as when—the retina hypersensitive—he sees the specter of some dear friend, in mere renewal of an old visual sensation: this is illusion. Hallucination, as we use it, is simply the externalizing of ideas.

2. Its history is a profound study in psychology.
It begins with the savage condition, and because it
pre-eminently characterizes childhood. One of the
many difficult lessons of childhood is to distinguish
between impressions from without and ideas within,
notions are so vivid : easily the rag baby becomes a
well-dressed living personality; readily the hobby horse
attains the size, grace, spirit and speed of a thorough-
bred ; and if the imagination be unusually active, the
child is in serious danger of becoming a gay romancer,
in time to be branded as an arrant liar.

The savage conditions representing the childhood
of the race is beset by the same peril. Savages externalize ideas and fill the world with their fancies; they
believe even their sleeping dreams. A savage dreams
of his friend, of his horse, of his dog, of the trees, the
landscape, the stars ; and he infers that not only friend
and enemy, but that animals, inert things, the moon
and the stars, have shadowy souls, and that these vaporous spirits actually come to him in his sleep.
Hence their almost universal belief in immortality,
and the pathetic custom of placing on the graves of
the dead weapons, utensils and food ; for the shades
of these things, it is supposed, will accompany the
soul of the buried into the land of shadows. Hence
also the slaughter of wives and slaves, horses and dogs,
over the burial place of a chieftain. So tenaciously
does this superstition, based on hallucination, persist
and push up even into low grades of civilization, that
down to 1781 the ancient funeral sacrifice of the war-
rior's horse was recognized at Treves by leading a dead
soldier's horse to his grave. A piece of money is still
put into the hands of a corpse at an Irish wake ; and

in most countries of Europe may still be seen the set-
ting out of offerings of food for the departed. All
this time-honored superstition arises from the externalizing of mere ideas.

From the same cause arise fetich worship, Nature worship, and in time an elaborate mythology.
The Fiji Islanders used to celebrate great sacrificial
feasts to their gods. These religious ceremonies, however, were mere orgies of gluttony, as all the animals
slain were greedily devoured; the deities were supposed to be satisfied with the souls of the departed
beasts. "In India," says Dubois, "a woman adores
her market basket, and offers sacrifice to it as well as
to the rice mill and other household implements. A
carpenter does like homage to his hatchet, adze and
other tools, and likewise offers sacrifice to them. A
Brahman does so to the style with which he is going
to write, a soldier to the arms he is to use in the field,
a mason to his trowel, and a farmer to his plow."
The worship of plants, animals, stones, water, wind,
sky, ocean, etc., is inevitable at an early stage of human culture.

3. While savages externalize their dreams after they
awake as well as before, even the most intelligent and
civilized do so at least during sleep. A very Plato, a
Shakespeare, a Darwin, must for the time fall under
the sway of his own vagrant fancies and believe them
real. The facts of dreaming, which we have fully
presented, are largely the working of this simple law
of hallucination.

The same is true of both natural and induced somnambulism, as we have observed: the mind, dreaming,
externalizes its visions.

4. Allied to the myth and dream fantasy of the mind, we must consider its romancing gift. The poets, novelists and artists have all been dreamers, their life work to produce in others a hallucination they voluntarily summon up in themselves; they see fictions and make them real, as landscape, as history, as personal beauty. All great art creators come to know their own handiwork by a kind of recognition, scarcely to be distinguished from actual acquaintance. And active minds respond in a kindred hallucination, joyfully self-imposed. A Ulysses, a King Arthur, a William Tell, a Pickwick, a St. Cecilia, a Venus de Milo, become as delightfully real to the imaginative, as though the legend, the novel, the painting or the statue were historical portrayal.

5. Hallucination is produced by certain narcotics, which occasion mental conditions varying from the profound quiet of perfect sleep to the most vivid dreaming or the most active somnambulism. These drugs paralyze the will, deaden the moral nature, confuse the reason and dull the senses, at the same time that they more or less excite the cerebro-spinal ganglia. If taken in doses appropriate to produce the effect, hallucination is inevitable. The narcotized person dreams, and, it may be, acts his dreams; but the dreaming is intensely spectacular, and the acting often bitterest tragedy.

Tobacco, though one of the least harmful of the narcotics usually abused, is yet noxious in every case and dangerous in many. Its first effect is to stimulate the faculties and soothe the feelings; its final result is to lessen mental power and enfeeble the will. It is said that no young man has graduated valedictorian at Harvard College who was an habitual user of this drug. Upon the young its action is peculiarly harmful.

De Quincey has portrayed vividly the deleterious effects of morphine on the same line of mental disease. He says: " Whatsoever things capable of being visually represented I did but think of in the darkness, immediately shaped themselves into phantoms of the eye; and by a process apparently no less inevitable, when once thus traced in faint and visionary colors, like writings in sympathetic ink, they were drawn out by the fierce chemistry of my dreams into insufferable splendor that fretted my heart." Besides these phantoms, projected against the darkness, there was a dream life of marvelous intensity. " Under the connecting feeling of tropical heats and vertical sunlights, I brought together all creatures, beasts, birds and reptiles, all trees and plants, all usages and appearances that are found in tropical regions, and assembled them together in China and Hindostan. From kindred feelings I soon brought Egypt and all her gods under the same law. I was stared at, hooted at, grinned at by monkeys, by paroquets, by cockatoos. I ran into pagodas, and was fixed for centuries at the summit or in secret rooms. I was the idol, I was the priest, I was worshiped, I was sacrificed. I fled from the wrath of Brahma through all the forests of Asia. . . . I was buried for a thousand years in stone coffins with mummies and sphinxes, in narrow chambers at the heart of eternal pyramids. I was kissed with cancerous kisses by crocodiles, and lay confounded with unutterably slimy things among reeds and Nilotic mud."

The fantasy of alcoholic intoxication, in time culminating in the horrors of delirium tremens, is known to all. In its first stages the ganglia narcotized are stimulated, and there are dancing, laughter and chatter. Later, the mind begins to externalize its thronging ideas, and the muscles to succumb to the secondary stupefying effect of the poison. Now the inebriate staggers, unable to co-ordinate movements perfectly, and hallucination becomes a prominent symptom and may rise into mania. This is the dangerous stage, when victims become suicides and murderers. Finally, a lethargy ends all psychic phenomena. In the trembling delirium hallucination is the principal symptom.

6. Fevers, in like manner and degree, exciting the nerve cells of the brain, through a poison generated in the blood or through mere hyperæmia, produce similar results.

7. Madness caps the climax by persistent, intense and tragic externalizing of ideas, especially that form which is called intellectual insanity. Here hallucination is the chief symptom, and the phenomena those generally pertaining to incoherent and troubled dreaming. Emotional insanity is less characterized by objectivity of ideas, and rather rouses to white heat states of feeling. Some one has said that the intellectually insane are furious dreamers, the emotionally mad dreaming furies; neither has self-control, nor is swayed effectually by judgment, reason or conscience, and both are prey to nervous disease.

Insanity presents no new facts; it gives us what we have abundantly in dreaming, hypnotism and the narcotic excitement, only more of it and for longer periods. It is chronic hallucination.

It is a very common affection. One of the authors's hobbies is, that every tenth person is insane and that every person is one tenth insane. That is, no one is perfectly mind-balanced, while many are seriously and up to the boundaries of chronic hallucination defective. Guiteau, who murdered President Garfield, though he seemed only a crank, in the autopsy displayed marked departures in the condition of his brain. Probably all cranks would be found, on postmortem examination, to have been afflicted with similar lackings or disease of central nerve tissue. Any one can become insane by extreme and prolonged activity of the mind or by brooding over evils; habits of health, laughter, sunshine, charity, patience and moderation are the preventives.

8. Hallucination is in itself a perfectly natural process; the dangerous element in it is its tendency to disturb the rightful balance of related intensities. We shall see later that sensations are more intense than perceptions, and these are more so than memories, while memories are more vivid than purely imaginative ideas; and that the mind distinguishes between these grades of mentality by a nice discrimination of intensities. Disturb the ratio, and the correct discrimination becomes impossible. It is easily disturbed in the child, because not yet fully established; and in the savage, because so far as subtler mental processes are concerned they are children. And with all men when outside impressions entirely fail, in dreaming no distinction can be made, and attention is rapt and deceived. Narcotics, fevers, and madness produce a like result by so overstimulating the ideational activity of the brain as to cause in the

intense vividness of the conception an outdazzling of the distinction. With an inebriated, delirious, or mad man outside reality compared with inner fantasy is but as a taper in the noonday sunshine. And poetry, fiction and other fine arts present the same unbalancing process, with delight self-induced.

9. Hence the great importance of guarding against our ideas. Sir William Hamilton well remarks: "Nothing is more dangerous to reason than the flights of the imagination, and nothing has been the occasion of more mistakes among philosophers. Men of bright fancies in this respect may be compared to those angels whom the Scriptures represent as covering their eyes with their wings."

CHAPTER XX.

HYSTERIA.

1. THIS distressing malady chiefly attacks young and nervous women, and is marked by outbursts of emotional excitement, convulsive bodily movements, hypoæsthesia and hyperæsthesia, and incomplete and transient paralysis. Intensely psychic in all its symptoms, it is marked by inordinate egotism, hilarity, depression, assumption of pretended diseases, catalepsy, trance and ecstasy.

2. It is a contagious affliction, and simulates the manner and method of the zymotic disorders. Carpenter's story of the mewing sisters will illustrate this point. The malady began in the hysterical tendency of a young girl to mew just as the clock struck nine and the morning session of the convent school was about to open. Soon other nervous girls caught the infection, and the solo became a chorus. Finally, all the young ladies without exception mewed, and the father was at his wits' end. One morning, however, he appeared with a horsewhip and anticipated the

stroke of nine by threatening to flog the first who should commence the concert; and his firmness was rewarded with silence.

The dancing mania of the Middle Age, named after St. Vitus, was a much more serious malady. Fits of nervous jactation, leaping and convulsion spread like cholera over Europe. Groups of temporary lunatics went whirling along the roads and through the city streets attracting the ill-balanced and spreading dismay. Those who came to look on and laugh stayed to dance and follow suit.

As nearly all intense religious experience predisposes to emotional excitement, it is not strange that with low-grade intelligence hysteria should accompany fanatical crusades, camp meetings and revivals; the remarkable fact is that the disease seems to be "catching"—one starts another, and a few excite many. During the early camp-meeting period of Kentucky, when that State was on the border and civilization raw, it was customary to plant stakes throughout the praying grounds for support of those who caught the "jerks."

3. But we are not to press the analogy of contagion too far. Some experimental psychologists of eminence plausibly advance the hypothesis that hysteria is not so much the disease of any organ as a general disturbance of nervous equilibrium. Says Myers: "Hysteria is not a lesion but a displacement; it is a withdrawal of certain nervous energies from the plane of the primary personality, but those energies still potentially subsist, and they can again be placed by proper management under the normal control."

Janet insists that no amount of hysterical disturbance, however prolonged or profound, need be regarded as incurable. At present the most approved medical treatment is by hypnosis and suggestion.

4. But better than cure is prevention. The hysterical should receive early training in self-control. Moreover, it must be borne in mind that such temperaments morbidly crave notice, sympathy and attention. With them to cause an excitement, to stir a thrilling sensation, is simply an overmastering passion. Hence the need of systematic repression and sympathetic neglect. When a " crisis " approaches, incredulity, indifference, contempt and even sarcasm are indicated. Indeed, it may be added that in chorea as well as in hysteria, and likewise in all purely nervous diseases, observation of symptoms and unnecessary sympathy aggravate the evil. Nervous movements should never be commented upon nor even noticed ; and if medically treated, it should be done, if possible, without the patient's learning the fact.

CHAPTER XXI.

CRIMINALITY.

1. TRACES of criminality may be found among animals—witness the " rogue elephant," which, once a member of some herd and now driven forth by the others, becomes morose, treacherous and murderous, much to be feared not only by men but also by its former associates.

2. Ellis defines criminality as a " failure to live up to the standard recognized as binding by the community. The criminal is an individual whose organiza-

tion makes it difficult or impossible to live in accordance with this standard, and easy to risk the penalties of acting antisocially."

3. As such organizations are sure to occur, criminality has ever been a marked characteristic of all social life among each race of men and in every age. No moral earnestness in any community, no political wisdom of any school of statesmen, has sufficed to eradicate these dark blots on human nature. Persistency and inevitability are their most perplexing features.

4. There are distinct kinds as well as many degrees of criminality. We have the criminal of passion, the occasional criminal, the habitual and the congenital. This variation depends upon the causes of aberration, and these causes are immediate and remote.

5. The immediate causes may be viewed psychologically or pathologically.

(1) Viewed psychologically, they are :

(*a*) Overmastering passions yoked with selfishness of disposition. This gives us the criminal of passion.

(*b*) A weak will and failure of principle. This gives us the criminal of occasion.

(*c*) When these two causes combine—strong passions and weak will—the habitual criminal results.

(2) Viewed pathologically, the causes are :

(*a*) Nerve defect. Nearly all criminals are deficient in sensitiveness of end organs, except in matters of sharpness of vision. In mental and moral endowment they are almost universally deficient.

(*b*) Nerve disease, occasioning unhealthy action of nerve centers. Hardened criminals are generally diseased with far-reaching constitutional ailments, and it is usually easy in *post mortems* to discover serious brain lesions.

6. The remote causes are :

(1) Atavism, or reversion to ancestral types. Ellis remarks that our own criminals frequently resemble in physical and psychical characters the normal individuals of a lower race. These abnormal natures are simply organizations out of date, that in an early savage state would have proved current as good citizens.

(2) Inherited virus, especially alcoholism, insanity and idiocy in parents. Indeed, criminality in this form is merely hereditary disease, which descends in moral scrofula, a horrid aptitude for evil-doing, from parent to child. We need only cite the case of the infamous Jukes family. The frightful story begins in the drunkenness, idleness and profligacy of one family, and continues through five subsequent generations, tracing the careers of seven hundred and nine of the twelve hundred descendants, who were for the most part criminals and prostitutes, vagabonds and paupers. Of all the men, not twenty were skilled workmen, and ten of these learned their trade in prison. One hundred and eighty received outdoor relief to an aggregate of eight hundred years recorded, and of (probably) twenty-three hundred years in all, at a cost to the public of one million dollars. Of the seven hundred and nine, seventy-six were criminals, committing one hundred and fifteen proved offenses. More than half of the women for six generations were notoriously unchaste.

(3) Failure of education, leaving the child the victim of idleness, poverty and contempt.

(4) Unfavorable environment, by which is meant

overcrowding, vicious and criminal surroundings, insufficient and unwholesome food, and all those conditions not sanitary which breed uncleanliness, immodesty, rudeness and disease.

(5) Excessive luxury, pampering the passions, encouraging selfishness, enfeebling self-control and diseasing the body, may bring about like results. Society has been not badly likened to a mug of beer, the froth on the top and the dregs at the bottom.

7. All criminality tends to assume an infectious and contagious character. The evil act of one dull nature shows to other dull natures a line of least resistance for criminal instinct, and one evil-doer begets many. Thus crimes often seem "catching"; they come in local visitations, and prevail in certain places as veritable epidemics, threatening the very existence of society.

It is quite a common occurrence for epidemics of suicide to break out in regiments of the French army, and it has become customary, on first symptoms of this mania, to remove the body of troops so afflicted to some distant region in order to divert attention and improve the general sanitary condition. Dr. F. W. Russell has stated that a lady patient of his, on reading a sensational newspaper account of the suicide of a drunkard, became possessed of the idea, and in a few days took her own life. Another, a friend of both parties, caught the dreadful contagion, and followed in the same awful course of folly. At least two cases, and probably three, attended by circumstances perfectly hideous, one in Australia and one in the West, have been directly traceable to the Whitechapel horrors, perpetrated by the wretch who signed himself "Jack the Ripper."

8. Where conditions favor, as in large cities, a criminal class—professional assailants of society, from whose ranks the prisons are regularly supplied, and who admire and emulate everything base because it is base—comes to the front.

9. There is also a literature of vice and crime, yellow-covered romances, dime novels, police gazettes, and daily newspapers, pandering to prurient imaginations, in minute description picturing the criminal as a hero, and furnishing to dull, vulgar minds the needed details for felonious action, all highly stimulant of brutal desires and purposes. We may be sure that, could the criminal items of the daily press be expunged, the rate of felonies would go down one half.

10. One of the most prominent features of the law-breaking temperament is its levity. The deep-planning villains of romance have no existence in real life. Milton's magnificent Satan is a mere poet's dream. Criminals are often cunning but never wise: thoughtless, illogical, the victims of a monstrous egotism, blown hither and thither by gusts of passion, they are lighter than vanity, their mental processes contemptible, their conclusions inconsequential, and all their conduct pervaded by an insane unreasonableness. They are seldom personally or mentally interesting, and generally dull, gross, repulsive and incapable.

11. The remedies are :

(1) Improved compulsory sanitation. We quarantine cholera, vaccinate against smallpox and maintain expensive boards of health. It has come to be viewed as a public duty to guarantee to citizens the conditions of bodily health ; and government is deemed society in its preservative and self-regulative functions.

It follows that we owe it to the poor and ignorant, and as well to the luxurious, to regulate their methods of life in the interests of a moral sanitation. To prevent overcrowding, debasing poverty and discontent among the poor, and no less enfeebling luxury among the rich, is the right and duty of society. Cleanliness, decency, sobriety, industry and self-restraint should be, and in time will be, enforced upon all citizens.

(2) Thorough education. The claim recently and often made that a large number of criminals are educated, is utterly groundless. Superintendent Brockway, of the Elmira Reformatory, declares that he has labored over criminals in prisons for over forty years, and can count on the fingers of one hand all the educated men he has found among them.

Education is a powerful preventive. Our public-school system should reach down lower, and take the very infant into its care on the plan of the public nurseries, and thus the day nursery for babes should be the primary department. The secondary should be the kindergarten for children, especially of the poor. Education should be compulsory, and when the little one enters what is now called the primary, it should be well along in private or public training. The child should graduate from the grammar school with something more than book learning, and rather in every wise prepared for useful grapple with the stern problem of life. In short, education should train healthfully all the various nerve centers, not only those of the cerebrum but also of the entire cerebro-spinal column.

(3) Social philanthropy, which should be personal, long-suffering and discriminating. Beggary can be

entirely repressed simply by an organized system of private relief, and at remarkably low cost, as is shown in the successful working of voluntary associations, called Union Relief, in New England.

Stealing and robbery disappear in proportion as wise and just laws regulating property are enacted and enforced.

And vice is very amenable to the influence of pure example and religious appeal.

12. The future of criminality can be prognosticated. It is not likely that atavism and occasional crime and the crime of passion will ever be eliminated. Inheritable virus, however, and habitual and professional wrong-doing ought to succumb to improved sanitation, wiser methods of education, and the elimination of those hard social laws which now work to reduce many to the level of the brutes.

13. One ought to distinguish between guilt and criminality. The former is an ethical, the latter a scientific, fact. Guilt may or may not be criminal; crime may or may not be guilty. Many a man who breaks no human law, and is adored of the community, has on his record "a damned spot" that all the waters of earth may not wash out; and many a hideous malefactor is innocent as the babe unborn.

Hence a just judgment of the criminal is exceedingly difficult. Several principles may, however, be applied safely.

(1) Depraved heredity is palliative and not cumulative of guilt. If the fathers ate sour grapes and the children's teeth are set on edge, so much the worse for the fathers; the guilt is theirs. The children are to be pitied; their freedom is thereby crippled.

(2) Unfavorable environment, also, is palliative of guilt. One is not worse because surroundings have been bad, but relatively better in the eye of justice. One's freedom has been compelled.

(3) Guilt concerns the exercise of the will, within its limitations and possibilities. Doubtless criminals are generally and greatly to blame; but we are prone to judge over harshly. "Let him that thinketh he standeth take heed lest he fall."

PART II.

MIND IN DETAIL.

SECTION I.

THE SENSORY AND MOTOR END ORGANS.

CHAPTER XXII.

THE EVOLUTION OF END ORGANS.

1. IRRITABILITY and contractility are primary functions of protoplasm, and hence in the lowest forms we have potentialities of that sense and motion found in the highest; for in the progress of evolution it is the irritability which expands into the nervous system and the contractility which is built up into the muscle machine. The two systems, the irritable and the contractile—that is, the sensory and the motor —develop in harmony and mutual dependence, the ready instruments of the ever - expanding psychic factor.

2. End organs are the results of this process of specialization, and hence are either of the irritable or of the contractile order—that is, either nervous or muscular, sensory or motor. At the very beginning

of specialization they may be both, but this only in the lowest forms. They are evolved to mediate between mind and both the outer and inner material world.

3. While protoplasm at its lowest does not see, hear, smell, nor taste, it is probably gifted with feeling, which we may conceive of as a dull, dim, perhaps only subconsciousness of itself and of the general effect upon it of environment. This primary feeling is the foreshadowing of sensation, and very likely similar to what in ourselves we name general sense.

Corresponding to this primary feeling is the voluntary action of rude forms, the mere self-change of position within a cell wall or in the open by a streaming of molecules or by elastic contraction and extension of the entire shape. The power of self-movement foreshadows all organs of action.

4. The earliest and rudest end organs are the false feet, lashes, whips and tentacles of protophytes and protozoa, already sufficiently described. These are at once irritable and contractile, of the nervous and of the muscular orders—that is, both sensory and motor—for they not only serve as limbs, they are also organs of touch. It must, however, be remembered that this primary touch is quite different from what we experience as such in ourselves. It is a crude forerunner of several senses, and as truly a prophecy of our smell and taste as of our tactile and pressure sensibilities; for its lowly possessors seek and find appropriate food, seemingly, by the aid of these simple members. Moreover, even our own smell and taste organs are but exquisitely delicate kinds of touch. The order of evolution undoubtedly was this touch of a rude sort,

followed by the same specialized, for heat, for smell, and later for taste.

5. The appearance of controlling or nerve cells must have given to this movement a great impetus. The sensory system separating from the motor underwent subdivision and elaboration. Temperature end organs must have come early, and have succeeded the general sense of hot and cold. Sight and hearing were at first less necessary, and received a comparatively late expansion. The muscular sense and muscular end organs were of course sequent to the development of the muscle machine.

6. The design of sensory end organs in high forms and low is to acquaint mind with its physical environment without the body and within. This sensory environment is the play of forces upon living matter— of light, heat, electricity, chemical affinity, gravity, pressure, etc. Sensory end organs are the effort of mind to interpret, counteract and master these forces.

7. This result is accomplished in all the higher forms by specializing superficial cells to receive one or another kind of stimulus, which cells in front elongate into a hair or thread and behind are connected with the sensory nerve centers. The apparatus is merely a hairlike process extending outward and connected by a sensitive cell with a nervous filament extending inward. The function of the exterior thread is to gather up the stimulus and convey it to the sensitive cell; in this the stimulus is converted into nerve force, which speeds to the brain, where the excitation becomes a sensation. Even the wonderful senses of man are only an elaboration of this simple structure,

which proves ample to meet all the urgent requirements of his complicated organization; and much
that we associate with them in thought is merely mechanical, only designed to bring to bear the stimulus
in the most effective way upon the sensitive thread.
Noses, ears, eyes and tongues are mere mechanical aids
to the vital operation of the concealed end organs.

8. The sensitive cells must be conceived of as
loaded with explosives, and their stimulation is a kind
of discharge, releasing energy. Some are fired by a
slight pressure, some by the molecular vibration of
odorous or sapid particles, some by undulations of air,
and others by the waves of an ethereal surf. Hence
we may classify the end organs according to the species of stimulus :

Mechanical........Touch.
Chemical..........Taste and smell.
Physical...........Sight, hearing, temperature.
Muscular ....Muscular sense.
Vital.............Vital sense.

9. It will be noticed that stimulus is in nearly all
cases vibratory. This is beyond question with sight,
temperature and hearing; but it is scarcely less certain with smell and taste, with the muscular and with
the vital senses. Upon this fact depends the quantity
and the quality of the impression. The form of the
vibratory curve determines its quality, its amplitude
the quantity.

10. General sense is described by Henle as "the
sum total or the not yet unraveled chaos of sensations
that from every point of the body are being incessantly transmitted to the sensorium." Weber defines
it as "an internal sensibility, an inward touch that

imparts information to the sensorium concerning the mechanical and chemico-organic state of the skin, the mucous and serous membranes, the viscera, the muscles and the articulated parts." Condillac called this "the basic feeling of existence."

It is by general feeling that we know our bodies as our own, and its sensibility forms the physical basis of personality. It gives us sense of comfort or discomfort, of *malaise* or healthful vigor.

If general sense have any end organs of its own, they must be very simple, and probably nothing more than terminating fibrils passing through minute sensitive cells, and which, starting out from the brain, end everywhere in the human body.

11. When the general sense is diseased or disturbed, hallucinations result, bearing chiefly upon the physical personality. A man imagines that he is two men, lying in two beds; or that he has long since died, and is but an inert thing. Esquirol describes a woman whose skin was completely insensible, and who believed that the devil had carried off her body. Ribot tells of a young man who, while maintaining that he had been dead for two years, expressed his perplexity in the following words: "I exist, but outside of real material life. Everything in me is mechanical, and takes place unconsciously."

CHAPTER XXIII.

THE END ORGANS OF TOUCH.

1. THESE, as we have seen, appeared early in the evolutionary movement, and are found in very low forms. With the latter they may subserve the purposes of the temperature and the smell senses. The hydroid polyps, the *Medusæ*, and the sea anemones have touch tentacles, usually arranged about the mouth. Sea urchins are equipped with touch rods and suctorial feet. Crustaceans and insects have touch hairs, often with a sensitive cell at the base. In the vertebrates the nervous apparatus is simply of naked fibrils lost amid the cells of the skin or of sensitive bulbs of connective tissue in which nerves terminate. In the great majority of fishes feeling is limited to the lips, to the fins and to special members called barbels. The tongue is the chief organ of touch in serpents and lizards. All reptiles that possess climbing powers develop the sense in their feet. Birds have touch papillæ on the soles of their feet to impart security of grasp.

2. In human beings touch is specialized solely for the appreciation of mass pressure; and the organs are numerous and in kind somewhat varied. These latter are located in the skin (integumentary or mucous), and consist of naked fibrils which end amid the cells or of fibrils ending in corpuscles.

The tactile corpuscles are of three kinds:

(1) Of Pacini, which are each a coating of many thin layers of connective tissue enveloping the termination of a medullated fiber, and are large ($\frac{1}{20}$ to $\frac{1}{6}$

inch in diameter). They occur most frequently in the palms of the hands and the soles of the feet.

(2) Of Krause, which are small capsules of connective tissue in whose center fibrils terminate in a coiled mass or a swollen extremity ($\frac{1}{250}$ to $\frac{1}{1000}$ inch). They are found in the conjunctiva in the tongue, the lips, etc.

(3) Of Wagner ($\frac{1}{300}$ to $\frac{1}{600}$ inch), situate in the papillæ of the skin. Within these, fibrils form two or three coils, and finally join together in loops; they are most numerous in the papillæ of the finger ends. Meisner counted four hundred papillæ in one fiftieth of a square inch on the third phalange of the index finger, and found Wagner's corpuscles in one hundred and eight of them.

3. The special function of the corpuscles has not been determined; it is, however, highly probable that the protecting connective tissue modifies the impulse of pressure received, so as to convey it to nerve terminations in a form better fitted for delicate and significant excitation than in case of naked fibrils. But they are not necessary for simple touch, and if no bulbar terminals occur in any part, still even there will be found sensitiveness to pressure.

4. The human body is covered with what are called pressure spots—that is, with areas marking the presence of some kind of touch organ. These are defined experimentally by pressing against the skin at every different point a sharp instrument. Light pressure excites a lively sensation, often accompanied by a sense of being tickled; heavy pressure arouses pain as if of a grain of sand forced into the surface. Between the spots the point will cause feeling of contact but not of

pressure. Undoubtedly fibrils terminate underneath the pressure spots.

5. The intensity and therefore the delicacy of the sense of touch depends upon the thickness of the cellular layer and the form and number of the papillæ. The two points of a pair of dividers can be distinguished by the tongue, if only one twenty-fourth inch apart; while on the cheek they may be one inch separate, and on the back three inches, and still give rise to only one impression. The tip of the tongue and ends of the fingers offer surfaces crowded with touch organs and highly sensitive to pressure. There are men in whom this gift is so exquisitely discriminating that they can tell simply by feeling the make and grade of flour. Sleight-of-hand experts possess this tactile dexterity to a very remarkable degree.

6. Touch is one of the spatial senses. There is a field of touch as there is a field of vision. Those who are born blind, through touch have definite space conceptions; the three dimensions are correctly apprehended. Indeed, in case of congenital blindness, touch, aided by the muscular sense, receives emphasis and plays a much more important part in life than when sight can be relied upon, and to some extent replaces the latter.

Still, even as a spatial sense, touch is far more reliable in its resulting sensations, perceptions and judgments, for having the assistance of vision.

7. In the blind, in the hypnotized and sometimes in the normal, touch attains exquisite sensitiveness. A Swiss blind man, among a group of wood-carving peasants, learned to carve in wood faces of men and women with a marvelous accuracy.

CHAPTER XXIV.

MUSCULAR SENSE.

1. This phrase describes a certain sensibility connected with the muscles, the stimulus being of two kinds—either an innervation of some muscle or a resistance by environment to such innervation.

Hence muscular sense is either of movements or of resistance. It keeps mind *en rapport* with its motory apparatus and cognizant of the obstacles encountered by its own attempted effort.

It must be carefully distinguished from the senses of mere contact and of mere pressure. Bring your hand alongside a book and you are in contact with it. Lay your hand flat upon a table and place the book upon your palm, and you feel pressure. Now, if you will lift the hand and so raise the book, you shall have muscular sense of resistance, as well as muscular sense of the innervation necessary to accomplish the work.

Landry tells of a workman "whose fingers and hands were insensible to all contact, pain and temperature, but whose sense of muscular activity was everywhere alert. If I made him shut his eyes and placed a large object in his hand, he was astonished that he could not shut it; but his only idea was that there was some obstacle to the movement of his fingers. I secretly tied to his wrists a kilogramme weight; he thought some one was pulling him by the arm."

2. In sense of muscular movement, the stimulation is probably central and occasioned by acts of innervation in central nerve cells. Sense of resistance, how-

ever, receives its impressions through fibrils directed everywhere upon the muscles, and these are the true muscular end organs.

3. The muscular is one of the three spatial senses, by which we come to know the material universe; its impressions acquaint us with the density, hardness and elasticity of matter. Indeed, sight and touch would fail in much of their usefulness were they not supplemented by the motility of the hand and eye muscles and the nice discrimination by successive movements of position, amount of innervation and force of resistances.

These spatial properties of muscular sensibility are greatly enhanced in value by that keen judgment of durations which accompanies it. Thus we can estimate distances by the time it takes us to traverse them under allowance for the force of resistance.

4. The diseases of this kind of sensibility lead to curious results. Demeaux tells of a woman who, losing all sense of resistance, though retaining sense of innervation, could will muscular action, but neither could she know the nature of the actual movement nor could she even judge as to the position of her limbs. Persons whose muscles have been anæsthetized can not tell in what position their members are or have been placed. Carpenter describes the case of a woman whose sense of resistance in one arm was lost, while the power of innervation was retained; she could hold her baby only so long as she gazed steadily upon her arm, vision taking the place of the lost sense in giving the requisite guidance to the sense of effort.

5. Muscular sense can be educated and attain wonderful keenness and precision; hence the sleight-of-

10

hand conjurer, the acrobat, the equilibrist and the tumbler. In hypnosis it may become extremely alert and discriminating, as in the case of the lady who could select any one of twenty silver coins by poising them on a finger and so weighing them.

CHAPTER XXV.

THE END ORGANS OF SMELL.

1. Odors are of value to the psychic factor thereby to gain acquaintance with environment, because material substances all readily give off superficial molecules and vary greatly in chemical quality. Odor end organs need to be specialized for reception of impressions conveyed by the chemical or physical energies of such molecules. How these energies are delivered so as to discharge the loaded sensitive cell, science has not yet explained; there is, however, little doubt that the impact is vibratory, and either of ethereal waves of low rapidity or rhythmic movement peculiar to the molecule in question.

2. While something answering to the sense of smell must be possessed by even the lowest animals, the most simple apparatus for this purpose is found in the *Medusæ*, in which we discover pitlike depressions lined with ciliated epithelium and supposed to be olfactory organs. Insects are abundantly provided with sensory hairs, knobs and cones on their antennæ, often numbering many thousands, which evidently are olfactory; for if you amputate these structures or coat them with paraffin, the result is complete obtuseness to smell. A

similar equipment of olfactory hairs and tufts of hairs is found with the mollusks; though snails, forming an exception, seem to smell—in part at least—with their horns. In fishes we have a highly vascular folded membrane, covered by cilia, lining one or two pits; and most fishes are attracted to bait not by sight but by smell. Amphibians have paired internal cavities. In birds the external nostrils are simple perforations, but in the cavities the sensitive surface is increased by projections and folds. In the mammals the olfactory surface is enormously increased by a bony labyrinth, carved into projections and depressions and covered with sensitive membrane.

The simple idea which is gradually elaborated in this series—the "motive," as a musician might say—is that of a single fibril exposed cautiously to contact with infinitesimal particles of substances floating on the air or dissolved in water. In each of the thousands of olfactory hairs on the antennæ of a bee there ends a fibril; while in olfactory cavities of more elaborate forms few or many such fibrils end in exposure. With the highest animals the structure, though still simple in its details, is extremely elaborate in multiplication of fibrils and provision for air passages.

3. In man the olfactory organs do not materially depart from those of his class, and are situated in the upper region of the nasal cavity, where there is an expansion. In this expansion there is a bony labyrinth lined with a mucous membrane that is abundantly supplied with sensitive fibrils. Two cables from the brain—the first pair of cranial—divide into fibers, the fibrils of which end in sensitive cells; these latter—spindle-shaped or columnar, with large nucleus

—penetrate the epithelial layer, and on the outer surface terminate in threadlike processes.

Smelling is excited chiefly on inspiration, and to
be keen the air must be breathed in deep draughts;
and snuffing, by creating a partial vacuum in the nasal
cavity and so increasing the amount of air drawn into
the olfactory region, intensifies the sensation. The
atmosphere had best be damp, and the mucous membrane concerned should be moist. When the threads
are well covered with odorous matter or coated with
mucus, as during an attack of catarrh, or when the
nose is filled even with an odorous liquid, the sensation ceases.

4. Smell is excited by exceedingly minute particles
of matter in the gaseous or vaporous condition floating
in the air or the same dissolved in water. A grain of
musk will scent an apartment for years, and at the end
of that time no appreciable loss of weight can be detected. This accounts for the extraordinary acuteness of smelling in certain animals, as the dog or the
deer.

Occasional instances of acuteness in men hint of
the inexhaustible possibilities of the human nose, and
lead us to infer that the only reason we do not enjoy a
sense of smell as keen and varied as our sight is the
fact that human exigencies of life and growth have
not required it. James Mitchell, born blind, deaf and
dumb, chiefly depended on smell for keeping up connection with the outer world; he readily observed the
presence of a stranger in the room, and formed his
opinions of persons apparently from their characteristic smells. Relatively imperfect as our organs are, it
is said that $\frac{1}{2000000}$ of a milligramme of alcoholic ex-

tract of musk and $\frac{1}{48000000000}$ of a milligramme of mercaptan can be perceived, while a current of air containing $\frac{1}{200000}$ of vapor of bromine excites a strong, unpleasant sensation. Humboldt declared of the Peruvian Americans that on the darkest night they could not merely perceive through smell the approach of a distant stranger, but could say whether he were Indian, negro or European.

5. In animals, and to some extent in man, smell conveys knowledge as to direction of the exciting cause. Put your finger in water occupied by leeches and they seek it. Fishes will thus find bait they can not see. In man this peculiar gift usually lies dormant; but that it exists appears in Braid's account of a lady who when hypnotized was so acute in smelling that she could, though blindfolded and at a distance of forty-six feet, follow a rose just as surely as a hound does a hare.

6. The olfactory sense may prove a source of deception. When covered with mucus in catarrh the nerve endings fail to respond to the strongest stimulus, and in certain diseased conditions they send false impressions to the brain—as when during a cold the author for several days smelt smoke and went about the house seeking fire.

7. Science has not reduced to definite mathematical, physical or even chemical relations the infinite possible varieties of olfactory impressions. As Wundt declares, these possess a " discrete manifoldness " which has an unknown arrangement.

CHAPTER XXVI.

THE END ORGANS OF TASTE.

1. TASTE and smell are allied, and were originally indistinguishable ; and now in all low aquatic forms the difference is hardly, if at all, discernible. Still, some think they discover taste organs in fishes, and Morgan is confident that they can be found in the maxillæ and probosces of insects, in minute pits supplied each with a taste hair. It is, however, more probable that taste organs have been recent in evolutionary history. First in the mammals do we discover clear evidence of a special apparatus, and even with these the sense of taste is introductory. Man himself in this regard is but where insects are in the matter of eyes, or where fishes seem to be in the matter of ears.

2. The taste organs in man are situated principally in the tongue. On the upper surface of the root of the tongue are found large papillæ, called circumvallate ; while on the tip and lateral margins may be seen other papillæ named fungiform. In the epithelial lining of these bodies, at the sides of the circumvallate, and at the sides and on the upper surfaces of the fungiform, are found gustatory buds. These open outward, and are $\frac{1}{300}$ to $\frac{1}{800}$ inch in diameter. Shaped like a Florence flask, they are composed of two sets of cells—an outer, which lines the organ, and is made up of nucleated fusiform elements bent inward like the staves of a barrel, and an inner group, five to ten, also of nucleated cells, each pointed at the opening and branched below. The branched lower

ends of the latter are continuous with nerve fibrils from the gustatory cable.

3. Taste is excited by soluble substances only, as the matter perceived must be minute enough to work its way into the terminal pores of the buds. Insoluble substances excite on the tongue only feelings of touch and temperature. Hence dryness of the mouth will lessen the sensation by preventing solution, and the neighboring secreting glands form an important part of the entire apparatus.

4. We may therefore surmise that the stimulus is molecular in action, the motion of impact being either an ethereal vibration or some rhythmic physical movement.

5. Taste seems a much more important function than it is, because constantly confused with touch and smell. A large part of what we call the taste of anything is its "feel" and its odor. A Shah of Persia once rebuked some Europeans for eating with knives and forks; he declared that the sense of taste began in the finger-tips (Höffding). Blindfold the eyes and close the nose, and a slice of onion on the tongue will not be distinguished from a slice of apple. We all enjoy vanilla flavoring, but only the odor is perceived. When smell and touch are rigidly excluded, there remain as kinds of taste impressions only sweet, bitter, acid and saline. In short, the sense is incipient: it is clear that the exigencies of animal existence have not been such as to evolve its possibilities. There is nothing absolutely to forbid the gradual future multiplication of kinds of taste impression.

6. The intensity of gustatory impression depends upon the number of buds excited and the concentra-

tion of the solution, not to speak of attention. But with all conditions favorable, the sense seldom is keen. It is, however, occasionally so, as in the case of Valentin, who detected bitter in $\frac{1}{1000000}$ of a solution of quinine. Moreover, it can be educated, as the nice discriminations of the professional tea-tasters show; though in this case the dexterity, after all, is largely of the olfactory sense. In subconscious conditions it is also often abnormally acute.

7. Like all the other senses, it may delude. Galvanic stimulation of the tongue simulates food impressions. A draft of cool air excites on the tongue the taste of saltpeter. Acetate of lead may be mistaken for sugar.

CHAPTER XXVII.

THE TEMPERATURE END ORGANS.

1. MIND perceives heat impressions through nerve fibrils terminating in the skin and mucous surfaces. The stimulus is to be sought in the invisible ethereal vibrations or heat rays which occur at the ultra-red end of the solar spectrum. The impact of this ethereal surf upon the fibril cells occasions that discharge of loaded energies which constitutes temperature impression.

2. These fibrils occasion corresponding cold and hot spots, which are minute and very numerous. The cold spots are sensitive to low, the hot to high temperature. Some parts of the human body are more plentifully supplied with the one than with the other. Thus the forehead and the back between the shoulders are ex-

tremely sensitive to cold, but only moderately so to heat, while the hands are equally excited by both conditions.

3. The fibrils are telæsthetic, and may be influenced by a body radiating heat from afar, even though as distant as the sun. You may prove this by holding your hand near a hot stove, but the while protecting it by a dense screen. Remove the screen, and in the fraction of a second, and ere the heat of the hand is appreciably raised, there will be a strong stimulation of temperature end organs.

4. There is a zero point at which fibrils produce no sensations. When they receive heat waves of any degree above this, the hot spots transmit inward the excitation of warmth; when the heat waves fall below, the cold spots transmit the excitation of coldness.

5. This zero point is variable; it changes for different parts of the body, according as they are or are not exposed, according as they are or are not well supplied with arterial blood, pursuant also to variations in the temperature of the air or of other bodies in contact. Thus it is higher in summer than in winter, in a hot room than in a cold one, etc.

The adjustment of the zero point to surroundings depends, of course, upon the evaporation of perspiration and the circulation of the blood, but also upon a certain power of accommodation. Plunge the hand into warm water, and having kept it there a moment put it into still warmer; this latter will seem warm only until the zero point is adjusted to the new conditions. Then, if the hand be returned to the first basin, the water in this will seem cold, though but a few moments before it gave quite the contrary impression.

6. The fibril cells are most sensitive to changes lying near their own zero point. Intense cold and heat do not occasion impressions strong in proportion to the intensity; indeed, overvigorous excitation, if prolonged, even reduces sensitiveness to slighter variations.

7. Delicacy of discrimination depends upon the locality and extent of surfaces involved. Thus water, in which the whole hand is immersed, seems warmer than some of a higher temperature into which only one finger has been plunged.

There is a limit, however, to this nicety of judgment; as the heat rays of the solar spectrum give us no such discernment of qualities as that afforded by the light rays, we simply feel gradations of quantity. The heat sense lacks what in other senses we mean by color, pitch, savor, etc.

CHAPTER XXVIII.

SIGHT.

1. THE light organs receive impressions from the luminous portion of the spectrum, and are adapted to convert the ethereal surf that beats against the human body—hundreds of billions of waves a second—into nerve force, to travel to the nerve center for final appreciation by the psychic factor.

Nature's method of accomplishing this great feat has been progressive; hence—

2. An evolutionary history. First came a general sensitiveness to light not localized. Among plants the

desmids have a light sense, and are able thereby to find the sunshine, whether by some special organ or by general sensibility we can not yet say. The earthworm is distinctly sensitive to light, and can even distinguish between colors, though quite eyeless, preferring red to green, and green to blue. The same is true of the blind proteus of the grottoes of Carniola. Some animals provided with eyes—the newt, for example—can distinguish between light and darkness by the general surface of the skin.

The next stage is of pigment spots. Certain plants in the motile form possess pigment spots as the organs of a light sense (as *Pandorina*). These spots are the rude beginnings of eyes. Low animal forms also have such pigment spots to serve the same end of light-seeking. *Euglena viridis* is a case in point.

Eyes proper began in eye-specks—in the worms, *Medusæ*, etc. These are simply expansions of an optic nerve into a brush of fibrils, which are fronted by a transparent medium, the whole shut in by a rudimentary lens. Eye-specks afford only a luminous impression, without distinct vision.

More fully developed are the ocelli or single eyes of spiders and kindred insects; these are endowed with a lens, a transparent medium back of it, optic fibrils, a layer of pigment and optic ganglia; and they no doubt afford something like true vision.

In the insects, and also in the crustaceans, single eyes—a great many of them, often thousands—are compressed and combined into compound eyes, which consist of transparent conelike bodies, arranged in a radiate manner against the inner surface of the cornea, with which their bases are united, while their apices are connected with the ends of the opposite fibrils. Vision in this case gives a distinct image of the field in mosaic.

In the vertebrates we find large, single eyes, in which the optical rivals the nerve complexity to provide perfect organs of sight. An eye, so perfected, is a dark chamber with a self-adjusting lens and a sensitive nervous screen. There is time here to dwell only upon the retina or screen, where the real end organs of vision are located, though the entire apparatus of diaphragm, ciliary muscles, muscles of accommodation, eyebrows, eyelids, lachrymal glands, muscles of the eyeball, etc., might well occupy many hours of patient consideration.

3. The retina or sensitive screen is the terminal membranous expansion of the optic nerve within the globe of the eye; it consists of nerve cells and fibers imbedded in a spongy, supporting connective tissue. It is the inner tunic of the orb, and is composed of no less than ten different layers. It begins on the inner surface of the choroid, with a mosaic pavement of pigment cells; resting on these is a layer of rods and cones, more than one hundred millions to the square inch. In front of the rods and cones are successive strata of nerve elements—fibers, nuclei and cells— connected with them. Foremost of all are multipolar nucleated nerve cells, joined by a network of fibrils to the optic cable, which enters the retina at a point of slight projection, near the center of the posterior hemispherical surface on the nasal side. The entire thickness of the retina is one thirtieth of an inch. At the center of the posterior hemispherical surface is a depressed yellow spot of superior sensitiveness, which is

the place of clearest vision, the organ of visual attention. The neighboring elevation, where the optic nerve penetrates the retina to distribute itself over the inner surface, is, on the contrary, devoid of visual elements and totally blind. It is believed that the eye distinguishes all the colors of the spectrum only at the yellow spot, which, in consequence, is termed trichromatic—that is, sensitive to the three primary colors and all their combinations. Not far from the yellow spot the retina becomes only bichromatic, and is green blind; while on the periphery color is entirely indistinguishable, and only light and shade are observed.

The yellow spot has within itself an area yet more restricted of most acute sensibility, only $\frac{1}{120}$ of an inch in diameter, containing no less than two thousand cones.

4. Though the mechanics of vision is perfectly clear, its physiology is not well understood. The formation of the image on the retina is in accordance with well-demonstrated properties of light; but how the retina converts each point of light into a nerve impression, to speed its way to the brain, we do not know. The fibers and cells of the retina are themselves indifferent to light stimulation, unless of dangerous intensity. The rods and cones, in connection with the pigment cells, receive and register the ether waves, but just how has not yet been discovered. The process seems to be photo-chemical. The fibrils receive their stimulus surely not from the light directly, but from the layer of rods and cones. The yellow spot is most sensitive, because here the cones are most numerous and delicate, while the blind spot is entirely

insensitive, because it fails utterly in rods and cones. An image large enough to cover one cone only is visible, and Lockyer claims that the color of a star throwing such an image is discernible.

5. A line drawn from the yellow spot to the center of the pupil forms the axis of the eye, and gives the direction of perfect vision. As the yellow spot is the organ of attention, and therefore of research and discovery, it is the only part of the retina that can be said to examine anything. The mechanical contrivances of the eyes are largely designed to bring the object of visual attention upon this most sensitive portion of the screen.

6. The retinal images are small—only about $\frac{1}{225}$ of the surface area of the object, at nine inches from the eye. The arc they subtend, the center of the lens being the center of the circle, is called the angle of vision.

7. The minimum limit of vision is conditioned by the distance of the retinal elements one from the other. Two stars can not be distinguished by the naked eye if nearer together than sixty seconds; this corresponds to a visual angle whose arc subtends the least distance between the cones in the yellow spot. A line not much more than subtending this angle appears uneven and knotted, because it falls at points on only parts of retinal elements, and lines of less diameter are not seen at all. Glass can be spun so fine as not to be seen even when magnified by the utmost powers of the microscope, and parallel lines can be drawn on glass that before all our efforts remain quite indistinguishable.

8. The impression made by light upon the retina

not only remains during the time of stimulation, but afterward for about one eighth of a second; so that two luminous impressions no farther apart than this interval appear continuous as one. A pin-wheel lighted and rapidly revolving appears as an unbroken circle of fire.

9. The excitability of the retina is soon exhausted: a bright light presently renders the part aroused temporarily insensitive. If the bright light be of one color, the part excited becomes insensitive to that color, but not to other rays of the spectrum. Look at a bright red cloth intently, and if the eyes be suddenly averted to a white surface a greenish spot will appear; in this case the capacity to see red is weakened, and only its complementary color in the white is perceived.

10. In some persons the necessary apparatus for discriminating colors accurately is lacking: they see no red or no green or no yellow.

11. Illusory impressions may be made on the retina. Press the closed eyeball on one side, and a light image appears on a dark ground. A blow will cause one to "see stars." Electrical stimulation induces light impressions of various sorts and degrees. The eyes are therefore the seats of possible illusions, and may become to the mind the sources of serious delusion.

12. Vision, through individual or ancestral education, can be brought up to a high degree of acuteness, and the same result may temporarily be secured by hypnosis. Jackdaws will perceive a hawk and show alarm when the sky is perfectly clear to human eyes. The author knew a young girl who possessed talent

for the painting of minute subjects so elaborate in detail that the result could be magnified four diameters without suffering in proportion or color.

CHAPTER XXIX.

HEARING.

1. THE end organs of hearing are constructed to acquaint mind with its environment by means of sounds, and the sensitive cells are discharged by undulations proceeding from sonorous bodies through elastic media. As sound is occasioned by the vibrations of bodies which, because they can vibrate, are called sonorous, and is transmitted in undulations through suitable media—air, wood, water, or any elastic substance—we may expect to find the terminal filament, in this case, an elastic hair.

2. The evolutionary history has been very striking. *Loxodes rostrum*, a beautiful ciliated infusory, exhibits along the back a row of small auditory vesicles, which probably afford a general sense of undulation, without discernment of pitch and timbre.

The *Medusæ*, in connection with their double ring of nerve matter, possess sense organs, which function, in some species, as eyes and in others as ears; in the latter case, projecting tentacles are furnished with otoliths and vibratory hairs. In some species the tentacle lies in a vesicle imbedded in the gelatinous substance of the disk and close to its edge.

Among invertebrates, auditory organs are very promiscuously located: in the foot of bivalves, in the antennules of lobsters, the forelegs of crickets and ants, the abdomen of locusts, the balancers of flies, the tail of *Mysis.* These generally involve one or more sacs, with otoliths and vibratory hairs. Sonorous vibrations are communicated to the sac either directly through hard parts or by a membrane exposed to the surrounding medium. If vibratory hairs be present, pitch is perceived, otherwise only intensity. Hensen, through a microscope, watched the two auditory sacs in the tail of a mysis, while a musical scale near at hand was sounded, and he found that special hairs responded to particular notes. When a note was sounded, the corresponding hair was thrown into such violent vibration as to disappear.

Among the vertebrates, the organ becomes increasingly complicated with evolution into higher forms.

In the lower fishes there is a simple sac; then, there is a sac—now called a vestibule—and a semicircular canal, each of which, filled with lymph and otoliths, receives filaments of the auditory nerve. In the lamprey there are two semicircular canals. In the higher fishes there are three semicircular canals, and the vestibule enlarges into a double sac. In amphibians, reptiles and birds there are always three canals, and conjoined to these appears a new sac called the cochlea; there is also increasing perfection of apparatus, in middle and external ears, for bringing to bear effectively the atmospheric undulations.

3. Omitting all minute description of mechanical contrivance, the internal ear of man is composed of two membranous sacs, filled with lymph and floating in lymph, inclosed in cavities of hard bone forming part of the skull. These sacs are connected by open-

11

ings with one another, and by membranes with the middle ear, and in their interior contain small, mobile, hard bodies—the otoliths. The membranous cochlea, which is to hearing what the retina is to sight, is a double tube wound about a spiral bone. On its inner surface, extending into the lymph, is a most remarkable series of threads, called the fibers of Corti, connecting tufts of hairs with fibrils of the auditory nerve; there are about three thousand of these, each with its tuft of hairs, and they are very generally believed to form a keyboard, and to be a musical instrument capable of responding to the utmost niceties of pitch and timbre. Whether the fibers vibrate, or the connected hairs, has not been decided; it is, however, probable that the hairs vibrate, and that the fibers are the converting sensitive cells.

4. As to the functions of the different parts, it is undoubtedly safe to say that the outer ear conveys sound vibrations to the membrane of the drum, whose throbbings are passed on by three little bones (ossicles—the mallet, anvil and stirrup) to the membrane of the oval window, which, itself pulsating, sets the lymph of the labyrinth into rhythmic motion; this rhythm throws into undulation the lymph of vestibule and cochlea, which breaks upon the nerve endings like the sea on a pebbly beach, increasing the intensity of their effect by lifting and dropping the otoliths, just as a surf lifts and drops sand and shingle. The last vibrations in the series are those of the elastic hairs, which sensitive cells convert into nerve energy and send to the brain.

It must be remembered that sound waves may reach the inner ear, though very imperfectly, through the bony parts of the skull and through the Eustachian tubes.

5. An auditory sensation lasts a short time after the cessation of the exciting cause ; hence, if sounds follow one another with sufficient rapidity, they appear as one and continuous. There must, however, be at least thirty per second to secure perfect continuity of impression. If these successive sounds be caused by regular and periodic impressions they form musical tones. There must be not less than thirty nor more than twenty thousand per second to insure perception.

6. Four qualities are discernible in musical tones—intensity, pitch, timbre and harmony.

Intensity depends upon the amplitude of the wave.

Pitch depends upon the form of the wave, and hence on the length of time in which a single vibration is executed or the number of vibrations per second. Acute or high tone is produced by rapidly succeeding vibrations, grave or low tone by very slow vibrations. The six or seven octaves of a piano cover from forty to four thousand per second.

Timbre or quality is that peculiar characteristic of a musical sound by which we may identify it as proceeding from a particular instrument or from a particular human voice. "It depends upon the number and intensity of other tones, called harmonic or partial tones, added to the fundamental tone" (Ladd).

Harmony describes the fact that several notes reaching the ear at once may, if the necessary relations exist between the numbers of vibrations, produce a sense of concord. Soprano, alto, tenor and bass unite to produce a pleasing effect, not from any arbitrary arrangement of things, but because of the application in musical composition of certain numerical laws of vibration.

7. The functional passivity of the ear favors its preponderating influence in generating mental characteristics. " The extreme case of the animal's control over the eye, and the absence of any control over the ear, made a difference in the degree in which the common animal appetites dominated the manner of the reception of the two kinds of impression. The passivity of the ear allowed auditory impressions to force themselves into consciousness in season and out of season, when they were interesting to the dominant desires of the animal and when they were not. These impressions got farther into consciousness, so to speak —before desire could examine their right of entrance— than was possible by impressions that could be annihilated by a wink or a turn of the head." Hence auditory communication of thought, and the enormous development of spoken language.

CHAPTER XXX.

THE MOTOR END ORGANS.

As a description of the various muscle machines constructed by Nature to serve as the apparatus of mind for voluntary response to stimulus would be tedious, and for our main purpose needless, we refer the reader for full treatment of the matter to the various manuals of anatomy and physiology.

—

CHAPTER XXXI.

SYNTHESIS OF SENSE IMPRESSIONS.

1. EVERY stimulation of an end organ, if it reach the brain, occasions a change in the central brain substance.

This change is accompanied by a corresponding psychic activity. Here we leap the chasm and pass from matter to mind, from force to thought.

2. If the stimulations of any one end organ are successive and the intervals brief, the psychic activity is not successive and with intervals, but continuous. The sensation of a tone, for instance, is not complex, though its occasioning stimulus be composed of thousands of sound waves.

3. If many end organs of the same kind be stimulated at once, the resulting impressions are co-ordinated—that is, they ultimately appear in consciousness as one color, or as one chord, or one landscape, somewhere or somehow connected and co-ordinated for that purpose. Take a landscape. Millions of retinal rods and cones are involved, and each has its message of color for the brain. On the retina the landscape is a mosaic of many elements, but in the brain it is a unity.

Each stimulation of color must be referred to that part of the field to which it belongs, and all must be combined. The result is simple, because a psychic process of synthesis has gone on. This result we may term a construct.

4. Moreover, if different kinds of end organs be stimulated at the same time, the resulting constructs may themselves be co-ordinated, and the final psychic result still seem quite simple; as when one suddenly grasps your hand, exciting both touch and muscular sense, or as when you eat chocolate, and touch, taste and smell are all at the same time aroused.

5. This work of noting, connoting and co-ordinating does not come into consciousness at all. It is a hereditary and instinctive gift, located in nerve apparatus, from which in the evolution of the nerve system consciousness has withdrawn. In its working we probably have a reminder of the psychical functions and mind activities of much lower stages of development than that of man. As it works in the dark, we can give account only of its causes and results. Treatises on metaphysics have generally ignored its existence and potencies, the importance of subconscious activity only lately having been recognized.

CHAPTER XXXII.

SENSATION.

1. A SENSATION is the psychic correlative of a synthesis of sense impressions. The synthesis acting as if it were simple, occasions in the brain substance a molecular rearrangement, and corresponding to this

molecular rearrangement occurs the psychic phenomenon of sensation. Of this stupendous correlation we can give no account. Well said Lotze : "All efforts to demonstrate how it comes about that the merely physical motion gradually passes over into sensation are wholly in vain. We must rather be satisfied with asserting that a necessity of Nature, which has hitherto wholly escaped our knowledge, has in fact united the two series of processes, the motions and the sensations —incomparable and irreducible to each other as they are—and has done this in such a way that a definite member of the one series always has for its consequent a definite member of the other."

2. Of the nature of the molecular rearrangement, also, we know nothing ; but we are permitted to infer that it is a change of great permanency, as sensations are nearly indelible brain records. This is proved by the memory phenomena of dreaming, somnambulism and hypnotism, to say nothing of the curious mnemonic experiences of those who are startled by accident. It is questionable whether anything short of organic brain disease, or the brain shriveling of old age, can erase the record of a sensation once become the property of the mind.

3. The process of sensation involves time—a brief period for the sensory impulse to reach the center, and a brief period for the psychic reaction in the center, which latter averages about one fifteenth of a second. Hence we need not be surprised that impressions entering very rapidly affect the mind as one continuous sensation. The cause is composite, the result simple— as when you rapidly swing around a lighted match in the dark and see only one ring of fire.

4. In order that impression result in sensation, it must contrast with what precedes. "The quite gradual increase in strength of an electric current will at length destroy a nerve subjected to its influence without any sign of sensation. By very gradual increase or decrease of temperature, a frog may be boiled or frozen to death without making the smallest movement. The pressure of air is noticed only when it varies. . . . There is no series of absolutely independent sensations, but every sensation is determined by the one experienced immediately before it or at the same time " (Hoeffding).

The necessary contrast may involve each or all of three differentia—quality, intensity and tone, all of which are relative.

5. The quality of a sensation depends first upon its complexity, and second upon the nature of the stimulus. Thus, in hearing, a sensation of simple tone differs from a sensation of harmony, owing to the varying complexities of the sense impressions, though all of a kind. And ether vibrations cause light or heat sensations, according as the length of the wave classes it in and above or below the red ; the stimulus, though in both cases vibratory and ethereal, varies greatly in its impact, and hence in its result.

6. Quantity (or degree) depends upon the strength of the impression. An impression must attain a certain intensity in order to occasion any sensation at all, or, in other words, a certain inertia of resistance must be overcome. This point is described as the threshold of sensation. The extreme maximum limit of perceptible impression is called the height of sensibility. In between lies the range.

Increase in quantity depends upon increase in the intensity of impression. But the ratio of increase for sensation is not the same as for stimulus—that is, if you would have more sensibility you must increase the stimulus not in the same ratio, but in a much greater ratio. If a stimulus s produces a sensation x, 4 s will not produce 4 x, but 3 x and 8 s only 4 x. This fact, formulated first by Weber, and later more accurately by Fechner, is generally stated by psychologists in these terms: "The strength of the stimulus must increase in geometrical progression in order that the sensation may increase in arithmetical progression." It must, however, be noticed that "Fechner's law" is only approximate; it holds for a medium range of sensations, and provided, in the cases compared, the attention be constant. Ribot's opinion is that it is "verified within certain limits for visual and auditory sensations, that it is contested for pressure, and does not hold for the other sensations." It must also be remembered that all sensations in threshold and height, as well as in intensity, are subject to considerable variation, due to physiological causes and personal temperament.

7. Moreover, the increasing intensity of a sensation is discontinuous. "A weight three must increase to at least four in order to give a new sensation of pressure; it gives no new sensation if only $3\frac{1}{4}$, $3\frac{1}{2}$, $3\frac{3}{4}$" (Lotze). No explanation is forthcoming.

8. The tone of a sensation—that is, its pleasurableness or painfulness—primarily depends upon its relative intensity and quality. Timeliness, heredity, habit, training, and intellectual, æsthetic, and moral appreciation, however, come in for a large share in deter-

mining both the relative quality and the relative intensity. Sensations have no absolute tone.

9. Sensations may not enter into consciousness, and in ninety-nine cases out of a hundred probably do not do so. They make a subconscious record, and, unless circumstances call them forth, never emerge from the deep into the sunshine. And of sensations that come into the purview of consciousness, by far the vast majority attract no attention and directly sink beneath the surface, as when one is passing over the country in a railroad train, listlessly gazing out upon the ever-varying landscape. To insure the notice of consciousness, and especially to attract its attention, a sensation must possess some degree of novelty or some measure of attractiveness. Familiar sights, sounds, tastes, etc., remain unnoticed ; if we feel them consciously at all it is to pass them by as a matter of course, unless they be of the sort to habitually appeal to love, vanity, fear, or some other strong motive. A savage often hears the roaring of wild beasts, but this is never received with inattention. A sweetheart often hears her praises sounded by her lover, but never with indifference. If, however, sensations be only habitual, monotonous, or stupid, they never venture to obtrude.

10. The freshness and vividness of sensations are intensified by attention. An absent-minded person, though a lover of music, may lose the pleasing effect of the most beautiful symphony or aria through sudden distraction of attention to some wonted train of thought. Either painful or pleasurable sensations may be dulled or quite ignored by persistent distraction. Consciousness turns the yellow spot of its mental eye upon the sensation, and it is seen more clearly.

Attention modifies the working of Weber's law to disturb the ever-varying ratio between stimulus and sensation in favor of the greater intensity of the latter. Moreover, it lessens reaction time.

11. Sensations may be illusory; they may not result from external stimulation, or they may not be normal. Quite frequently the end organ in a reversed reflex action is aroused by the brain, and the mind plays pranks on itself. The optic nerve quivers with a message not communicated by the ether waves, the olfactory signals an odor not on the breeze, the tongue tastes when no food is in the mouth, etc.

The simulation may be occasioned by disease of the end organ, as in deafness, when one hears bells ring and gongs sound and voices call, or as in a cold, when one smells nothing but imaginary smoke.

Many persons are subject to auditory spectra; they hear unaccountably music or words or their own name. Huxley says: "I know not if other persons are similarly troubled, but in reading books written by authors with whom I am acquainted I am always tormented by hearing the words pronounced in the exact way in which these persons would utter them, any trick or peculiarity of voice or gesture being also very accurately reproduced. And I suppose that every one must have been startled by the extreme distinctness with which his thoughts have embodied themselves in apparent voices."

Moreover, such illusions may be at will produced by artificial combinations of sensations. Ventriloquism is a good illustration of sensory illusion, deceiving the ear by simulated tones, and the eye by corresponding gestures. Optical illusions are very numerous, because

what we ordinarily consider simple visual sensations
are often complex aggregates not only of various sensations tactual as well as visual, but also of recollections and judgments. If a continuous series of pictures of one object be impressed upon one part of the
retina, the mind judges that they are due to a single
object undergoing changes. This is the principle of
the zoetrope.

It is commonly said that one must believe one's
senses; it is evident from the foregoing that this claim
fails of an absolute validity. In general true, it needs
careful delimitation; and much of the superstition that
has ever cursed the human mind has based itself upon
mere sensory illusions, through misinterpretation become delusions—voices from heaven, spectral music,
ghostly apparitions, and so on.

12. It must always be borne in mind that a sensation is utterly different from the things or motions
that cause it. The ear is excited by sound vibrations,
and the eye by light vibrations, but neither sound nor
sight is in the least like any series of vibrations.
Moreover, the same stimulus may excite different end
organs so as to produce sensations which shall not in
the least be like each other. If a man squeeze your
hand you feel his friendly touch; if he squeeze your
eye you "see stars." Electricity will occasion lumi-
nosity, taste, smell, or touch, according to its point of
attack. Different stimuli, on the contrary, exciting
the same end organs, may also occasion utterly different sensations. On the eye light produces vision, electricity a mere luminousness, heat only pain, and sound
no effect whatever.

13. Sensations may themselves blend together in

groups which seem simple, as when one listens to a
symphony or an oratorio. Here the sensation of hearing is composed of a vast number of sensations of
successive chords of music, and differing qualities of
voices and instruments.

14. Sensations form the ultimate material for
thought. "Systems about fact must plunge themselves into sensation, as bridges plunge their piers
into the rock. Sensations are the stable rock—the terminus *a quo* and the terminus *ad quem* of thought"
(James).

15. Pure sensations can only be realized in the
earliest days of life, when the babe's experience, again
to quote James, "leaves its unimaginable touch upon
the matter of the convolutions, and the next impression which a sense organ transmits produces a cerebral
reaction in which the awakened vestige of the last impression plays its part. . . . The complication goes on
increasing till the end of life, no two successive impressions falling on an identical brain, and no two
successive thoughts being exactly the same." Wundt
claims that "pure sensation is an abstraction, which
never actually occurs in consciousness"; he urges that
"every presentation (*Vorstellung*) is a synthesis of
a plurality of sensations."

16. Sensations develop affinities, behaving much
like the molecules of substances: as these associate
themselves in series and groupings called compounds,
so sensations spring into one another's arms, embrace,
join hands, and form series and groups. Even if they
enter in comparative isolation, as in case of the sudden
report of a gun, directly they associate themselves with
other and relevant sensations.

17. There is nothing, seemingly, to forbid the rise of new kinds of sensation, divaricating branches of those already possessed or based upon entirely novel end organs yet to arise. Some animals do apparently possess senses not enjoyed by man, and in man there are manifest gaps to fill, as a sense to interpret the ultraviolet rays of the spectrum, a magnetic and an electric sense. Indeed, within a few years able scientists have announced the discovery of a new series of end organs in the semicircular canals; which, commonly considered instruments for gauging the direction of sound, are now claimed for a sense of rotation.

CHAPTER XXXIII.

THE PERCEPTIVE PROCESS.

1. Is complex. We have been taught to name it perception, and assured that it was one of "the faculties"; but really it is a composite of many and different mental habitudes. It is the whole mind in the act of acquiring knowledge.

The characteristic feature, however, is the externalizing of sensations. We have seen that sensations are subjective. The perceptive process externalizes them. A solitary perception is an aggregate of sensations externalized. This is why Kant's famous dictum is true, and the mind does not know anything "in itself," but in its qualities.

2. Mind is equipped for the perceptive process by certain original and certain acquired gifts.

The original gifts are the ideas of time and space,

which seem to be necessary and universal forms of thought.

The acquired gifts are the mind's practical wisdom, the results of previous observation and experience, which latter are largely ancestral—that is, instinctive—knowledge crystallized in inherited brain structure. We receive as a bequest the accumulated practical wisdom of countless generations of sentient beings, and upon this depends the methods and accuracy of our perception. To this ancestral dexterity we gradually add the acquired dexterities of our own lifetime, beginning with early infancy.

But all this is under the pervasive reign of the ideas of time and space.

3. The evolutionary history of perception, could we know its true inwardness, would be one of the most fascinating chapters in psychology; but we may only surmise the story of that dawning knowledge of the world which gradually shone—more and more unto the perfect day—upon primeval mind, and in course of ages, in ever-expanding forms, approached man's intuition of the universe. No doubt it is, however, in the after-stages, at least dimly recapitulated to us in our own infancy and childhood.

The babe at first experiences sensations, but scarcely perceives. Soon, however, the sensations of light and sound, of warmth and touch, which at first were felt without recognition, begin to excite a responsive smile or cry, and the infant is joyfully observed to "take notice." From this point onward to manhood life is largely a training in perception.

4. The perceptive process is threefold : it localizes sensations in or on the body; it projects them into

space, attributing them to things; it arranges them, where they permit of it, in order of sequence or in spatial perspective.

To do all this it marshals sensation, memory, imagination, judgment and every mental habitude in its service.

(1) Localizing of sensations in different parts of the body is the result of observation and experience, partly ancestral and partly individual. That it is the nose which smells, the tongue which tastes, the eyes which see, and the ears which hear, is the inevitable conclusion of reasoning based upon touch and muscular perceptions. Thus to smell accurately one must sniff, and sniffing calls attention to the nose. We have good reason for believing that subconscious localization has become very precise and minute. Lotze supposes that every feeling point has acquired a "local sign" of its own, whereby it is distinguished at the nerve centers from all the others. The particular impressions at first are intensive merely, but the mind gives to them an extensive significance. This famous theory of "local signs" is Lotze's principal contribution to modern psychology. Or, to state it in the philosopher's own words: "Every impression of color, r —for example, red—produces on all places of the retina which it reaches the same sensation of redness. In addition, however, it produces on each of these different places, a, b, c, a certain accessory impression, a, β, γ, which is independent of the nature of the color seen, and dependent merely on the nature of the place excited. This second local impression would therefore be associated with every impression of color, r, in such manner that $r\,a$ signifies a red that acts on the point

a; r β signifies the same red in case it act upon the point b. These associated accessory impressions would accordingly render the soul the clew, by following which it transposes the same red, now to one, now to the other spot, or simultaneously to the different spots in the space intuited by it. The result is that we recognize the locality of an impression, provided we have already had experience at that point, and the more habitual the experience, so much the more accurate the recognition."

Fatigue, lowered temperature, or other abnormal condition may mar the precision of localization by obscuring the impression.

Disease also may work derangement.

(2) The projection of sensations is much more complicated, as these have no length, breadth, height, objectivity, or external reality. Even if localized by local signs, they are so far only intensive.

Projection grants them extension. "Objects are perceived in space as situated in a right line off the end of the nerve fibrils, which they irritate" (Ladd). Thus the retina receives the image of a landscape in points of light, exciting a vast number of retinal elements, which image is inverted and reversed; but in perception each point of light is referred back to its place of origin and given existence in space.

Hence we have a field of touch and a visual field.

The field of touch is the most ancient method of projection, and must have been acquired by the lowest protozoans at the very beginnings of life. This is the only field of space persons born blind possess.

The field of vision easily results from the projection of sensations received in images on the retina

12

into space, *seriatim*. This, in case of compound eyes,
is manifest; but it is as true with simple.

There is no field of smelling, tasting, hearing, etc.,
though sounds are in a measure localized, and odors,
savors, heat and cold are perceived under at least
ideas of space.

(3) The arrangement of projected sensations under
laws of extension and harmony is yet another function
of perception, and it may be tactual, visual or auditory. In the field of touch—direction, the three dimensions of extension, hardness, distance, size, etc.,
are noted. In the field of vision—color, direction, distance, perspective, size, etc., are noted.

Sounds are arranged as successive or contempo-
raneous in noises, tones, melodies, harmonies, etc.

Taste and smell, hot and cold, fail of arrangement
in dimensions, motions or series of any kind, either
because they are incapable of it or because the faculty
of perception has not yet been educated to detect the
laws that govern them.

5. In the field of touch, perception is determined
by pressures nicely varied, weighed and compared,
by muscular sense of resistances also nicely varied,
weighed and compared, by keen discrimination of the
degrees of muscular innervations required for movements, and also no doubt by judgments of temperature. Pressure, resistance, motion and heat are here
the prime factors of knowledge.

In the field of vision, perception depends upon the
muscular movements of the muscles of the eyeball and
of the lens, on the sense of innervation, and on the
differential impression of separate retinal elements.
It is materially improved by the motion of the eyeballs

and the conjoint vision of two eyes. Its results are corroborated by the aid of touch.

Of direction we learn by comparing the line connecting the yellow spot and the point of regard with general position of the body.

Form in two dimensions is perceived by the wandering of the point of regard over an object observed. Form in three dimensions results from the conjoint use of both eyes viewing the object from slightly different standpoints, the images falling on identical or corresponding spots of the retina, or from the use of the accommodative muscles of either eye.

Distance is gauged by the angle of muscular convergence for near objects, and by judgment of aërial perspective for those afar. The muscles of accommodation assist for near objects, and if trained will of themselves give accurate measurements.

The apparent size of an object depends upon the extent of the retinal image in connection with an estimate of distance. The degree of illumination avails also in forming an estimate. " Distances we estimate (very indefinitely) as smaller for bright objects, larger for dark ones—much more accurately as smaller so long as the interior delimitation of things continues to be clear, larger in case it makes a confused impression as a whole. We principally, however, employ three factors—the actual magnitude of a thing, its apparent magnitude, and the distance, in order from two of them to ascertain the third " (Lotze).

6. The result of the perceptive process is a percept or intuition of some thing, and how far the thing we intuit is identical in its qualities with our percept depends upon the accuracy of our previous experience

and observation and the keenness of our present judgment. It constitutes at best only a relative and imperfect knowledge, though we are not justified in calling it hallucination, as Taine and other experimental psychologists have done. Hallucination is the externalizing of ideas; the externalizing of sensations is knowledge. Of course it is knowledge of the world as we know it, the world of phenomena, of what appears to us and must vary much from the cognition of other and different creatures. A dog's world, or an ant's world, or a fish's world, could we somehow perceive with their end organs, would doubtless much amaze us, for both defects and revelations.

7. This is the more important, because here as everywhere illusion is possible. The most correct sensations may be faultily projected and arranged, or mere subjective feelings or ideas may be externalized as realities. I perceive smoke, and there is none. I perceive an absent friend who directly vanishes away into thin air. I hear my name called when no one spoke. I look out from the car window and perceive the landscape moving. These illusions are due to disease, to faulty judgment, feeble imagination, instinctive inferences, etc.; but they condition the accuracy of the perceptive process. They may also be owing to an overactive imagination; the same poet or dreamer who by normal hallucination makes real his own ideas, may, by a reverse process, but through the same cause, idealize realities. Wordsworth, speaking of his own boyhood, said : " I was often unable to think of external things as having external existence, and I communed with all I saw as something not apart from, but inherent in my own immaterial nature. Many times while

going to school have I grasped at a wall or tree to recall myself from this abyss of idealism to the reality.
At that time I was afraid of such processes. In later
times I have deplored, as we all have reason to do, a
subjugation of an opposite character, and have rejoiced
over these remembrances." Tennyson said one day :
"Sometimes, as I sit alone in this great room, I get
carried away out of sense and body and rapt into
mere existence, till the accidental touch or movement
of one of my own fingers is like a great shock and
blow, and brings the body back with a terrible
start."

Indeed, it is through discovery of our follies of perception that we become truly informed. Untold generations of sentient creatures have bequeathed to us
their practical wisdom, acquired through countless
mistakes, failures and miseries, and their bitter experience has suffered not in vain. On the whole, and in
broad averages, our intuitions of the universe are in
accordance with the facts. Our knowledge of things
is tentative, but real ; relative, but suitable ; finite,
but enough. Science has much to do to disabuse our
minds on many points, but still a "spade is a spade,"
and "a man's a man for a' that and a' that."

8. Moreover, the education of this perceptive process is still in active operation ; nay, we ourselves may
each educate perception as we do memory. Commence
teaching your children when only babes how to see
and hear everything, and how to judge things accurately, and long before they themselves open books
keen eyes and ears will have come for that book which
is always and everywhere open. This training can be
made minute. It will be remembered that the two

points of a pair of dividers can be distinguished as separate when only one millimetre apart on the tongue,
while on the back sixty-eight millimetres will but just
suffice to insure the same results. Now Volkmann has
shown that education by practice will fully double this
sensibility for any given part, and in a very brief period. The marvelous tactual perceptiveness acquired
by the blind affords further illustration.

9. The vigor of perception depends upon a concentration of attention upon the psychic action occasioning it. We may easily see and not perceive, or perceiving, not perceive clearly. That the process may
be keen and accurate, the mind must direct and super-
vise. You smell odors of flowers—you stop, sniff the
air, and perceive that it is mignonette. Or you hear
a bell, start up, and on the second stroke, listening,
perceive that it is the fire alarm. A steamer passes on
the river; you shade your eyes, look intently, and perceive the name on the pilot house.

10. Perception works as readily below consciousness as does sensation. Indeed, subconscious perception takes cognizance of a world of facts, which the
surface ego does not immediately, and may never recognize. Coleridge cites the case of a boy who, at the
age of four, suffered fracture of the skull, for which
he underwent the operation of the trepan. He was at
the time in a state of perfect stupor, and after his re-
covery retained no recollection either of the accident
or of the operation. At the age of twelve, however,
during the delirium of fever, he gave his mother an
account of the operation and the persons who were
present at it, with a correct description of their dress
and minute particulars.

Our entire treatment of the subconscious has already made this fact emphatically apparent.

A highly intelligent anonymous Englishwoman, who has devoted much attention to crystal vision, one day saw in her crystal a first-page column of the London Times and an announcement of the death of a lady at one time a frequent visitor in her circle, with date, place and circumstances. This startled her, and seemed clairvoyance; it was, however, only memory, for later she discovered that the announcement had been in a morning paper she had glanced over. It had stimulated the retina and been telegraphed to the brain, and there became a sensation subconsciously perceived; only, however, to sink into the limbo of percepts unheeded.

CHAPTER XXXIV.

MEMORY.

1. MEMORY is simply a name for the persistency of sensations and of other mental states; as Cicero called it, "Thesaurus omnium rerum."

It is probable that every mental state leaves some indelible record upon the brain. This we infer from the flashing into consciousness, at times of startling calamity, of a great number and variety of facts supposed to have been forgotten, and from the phenomena of hypnotism with its multiple mnemonic chains. But there are many other sources of information on the subject. Old age will often restore events of childhood seemingly utterly lost to mind, while disease in feverish excitement will bring to the front a vast congeries of facts which seemed hopelessly faded.

The author has observed that, though fine music gives him the greatest delight, he easily forgets the strain on a first hearing, and can not after the concert repeat a single melody; but if his mind dwell lovingly upon the brief joy and often return to the memorable event, after a week or two he is found humming to himself the identical lost airs: his mind, impressed with his desire to recall the strains, laboriously fishes up out of the depths of subconsciousness the seemingly lost but really indestructible series of sensations.

2. While we claim that memories are practically indelible, we must not be understood as asserting that they are unalterable. The physical apparatus of permanence may itself undergo changes involving considerable modification of the original sensations. If you cut your name into the bark of a tree it will remain there until the trunk decays, albeit for five hundred years; but the letters undergo gradual distortion with the swelling of girth and lapse of time; so sensations recorded may be distorted, fade in distinctness, and blur in outline, but without entire effacement. This fact forms the physical explanation of many mistakes and much untrustworthiness in human testimony; the original record may have been truthful, and it is indelible, but now weather-worn, time-eaten.

3. Forgetfulness is not loss of any part of the acquired wealth, but only the inability at the moment to find the particular item of treasure wanted. The memories of most persons are mere junk shops, and people acquire only to forget. A retentive, well-stored

memory is an orderly museum, where everything is labeled, catalogued and easily accessible.

4. Particular items of memory develop affinities. They tend to group themselves together, much as do sensations, and under the same laws of association. You never find one that does not suggest another. Hence we speak of memory's chain, and we may be sure the links, though they fall out of sight into a deep of subconsciousness, are never dropped.

5. The intensity of memory depends upon the original intensity of attention fixed upon the sensations or mental states restored. Facts that enter subconsciously, or that, unnoticed, are quickly dropped from consciousness, though they persist, do not easily come up into view, and are among the countless hosts of the forgotten. Hence the transitoriness of that learning which is described as cramming. As we have not well perceived anything which has not fallen upon the yellow spot of the eye, so we have not well memorized anything that has failed to fall upon the yellow spot of consciousness.

6. In old age, the failure of memory is a breakdown of the brain structure. The ganglia shrivel and lose weight, size and vitality. The last records go first. The physical apparatus may also be erased or confused by various forms of disease. Possibly this proves that groups of remembered items are, in their physiological records, localized in the brain. We are not, however, to suppose that each separate item is stored up in a separate nerve-cell, developed or appropriated for its indwelling. Rather all the facts bear in favor of the hypothesis that every nerve-cell, and indeed every cell in the human body, has its own mnemonic series.

7. The very extraordinary facts of dual or trinal consciousness show that memory admits of curious cleavages. A somnambulist recalls in the waking stage little or nothing of the sleep-life. A dual personality is a memory, crevassed into two; and sub-personalities are simply subtrains of consciousness, developing each a mnemonic series of its own.

CHAPTER XXXV.

THE RECOLLECTIVE PROCESS.

1. THIS is the mind's power to restore to conscious-ness former mental states. Memory only preserves; recollection revivifies.

The restoration is only an image or idea of what once was a sensation, percept, or other mental state—a fainter renewal of the former experience.

2. We call this a process because, like perception, it is very complex, involving the whole mind in use of many of its habitudes. It involves

(1) Personal identity—taking for granted that I am the same I that I was.

(2) Discovery. Recollection must search the archives of memory and find what is wanted—in most cases a Herculean task.

(3) Recognition. What is found must in the find-ing be recognized as the identical former state laid away.

(4) Restoration. This is the faint revival of the original state.

3. Recollection may work either automatically or volitionally. In the former case it is suggestion, in the latter reminiscence.

Suggestion is most active in reverie and the various forms of dreaming. Reminiscence is most active in those various forms of cerebration which involve mental effort. Suggestion is play, entailing but little waste of nerve tissue or force, as it follows lines of least resistance, and recalls only what is uppermost. Reminiscence is work, and often extremely hard.

4. The laws that control suggestion and reminiscence are those principles of association already stated, under caption of the Enchaining and Grouping Function of Consciousness—the laws of simultaneity and affinity. Memories cohere, owing to their co-existence or immediate succession in time; owing to their resemblance, contiguity, or integral or causal relation to one another. A memory comes up into consciousness or subconsciousness only as it is fished out of the depths by help of some chain of sequence in which it forms a link. But bear in mind that much of this chain will likely remain out of sight, and, furthermore, that the chain coheres in an infinitely tangled network with numberless other lines of association.

Dreams are immediately forgotten because isolated in the act of memorizing from all real fact and event. Con over your dream immediately on awaking, write it down and talk about it, and it will become easily accessible for future reference, because its affinities with reality have been developed.

5. Hence there may be skill in recollection, and indeed there is a science of mnemonics. He will recall his memories easily who, in storing them away, pur-

posely associates them with other easily accessible and
similar or contrasted facts, notions, or events.   It
needs only to bring things difficult to recall into touch
with things easily and often remembered.

The defect of learning by rote, or of committing
knowledge verbally to memory, is that the entering
series is only a single chain, and not, as it should be, a
network for each link.   Great is the confusion of
one who " parrots off " knowledge, if interrupted " *in
medias res* "; he must commence all over again.   His
recollection is a thread, and threads are easily broken,
and more often lost.

The cramming of knowledge is equally defective,
and for similar reasons.   The mind has no time or
strength of digestion for crude and confused heaps of
material.   Memories are allowed no opportunity to develop those affinities so useful in recollection.

Time and attention are both indispensable in the
real acquisition of knowledge.

6. The vigor of recollection depends upon a number of considerations—whether, for instance, the thing
to be recalled be vivid, whether it be pleasant, whether
it have been recalled a great many times, and whether
it have recently been in review.

Extraordinary vigor is of course a rare native gift.
Sir William Hamilton cites the remarkable case of a
young Italian to whom, in a party of distinguished
men, words were dictated, countless words,—Latin,
Greek, barbarous, significant and not significant, dis-
jointed and connected—until all were wearied but
the youth, who called for more.   Every word was repeated in its order without hesitation.   Then, com-
mencing with the last, he repeated them backward

until he came to the first. Then he repeated the first, third, fifth, etc., in any desired series. He claimed that he could do this with thirty thousand words, and even after a year's time. It is said of Grotius and of Pascal that they forgot nothing that they ever thought or read. Leibnitz and Euler could each repeat the whole of the Æneid.

7. We have seen that a memory may become blurred; we must add that recollection may falsify even a clear record. Many habitually exaggerate, understate or even distort items of memory presumably legible. They purpose no deception, and deceive themselves. This is a form of mental disease. Other defects of this faculty are clearly kinds of brain disease, and are very curious. Certain things it may be impossible to remember, and certain others impossible to forget. A man suffering from aphasia can not recall a particular letter, and always drops it out; another, brain fevered by remorse, can not put out of his sight the imploring face of his dying victim. The rememberings and forgettings of hypnotics will be thought of in this connection.

8. It is as important to learn to forget as to learn to remember. The vast majority of the sensations that enter the mind are trivial, vulgar, perhaps vile, and had best sink quickly to rise no more, down into the abyss of the subconscious. When Simonides offered to teach Themistocles the art of memory, he replied that he would rather learn to forget, "for I remember even that which I do not wish to remember, but can not forget what I wish to forget."

9. The accuracy of recollection may be unequal for different classes of facts. Thoughts are easily recalled,

but faces and names forgotten, and so on. Some remember best *verbatim*, and some best *ad sensum*. This is explained by idiosyncrasies of brain formation and psychic organization.

10. A recollection is always fainter than the mental state recalled. You glance upon a new, beautiful face, and it is seen vividly and with lingering gaze; but close the eyes, let all after-images fade, and now recall it. Alas, what a shadow of that glory in flesh and blood! And hence on, you shall find the image fade until you may no longer recall it and only certain things about it. But let another be pointed out as the very face in memory, and at once recollection rejects. It may not be able to show what the countenance is, but it can readily say what it is not. Manifestly in subconsciousness there is a more correct image preserved, than recollection can recall.

11. As we might expect, recollection takes much more time than the original state, if that have been a sensation, as this has not only to be revivified but first discovered and recognized. In other words, it takes longer to find a specimen in a museum than to place it on the shelves.

CHAPTER XXXVI.

IMAGINATION.

1. IMAGINATION is the mind's power to hold up before itself for study the mental states it has recalled.

By this function the mind idealizes its preceding states; while we perceive by sense what is present, we form an idea of what is absent.

Yet the idea is not so much a picture as a real revival of the sensory or motor elements of the thing ideated. It is not a sensation nor a percept, but still a true sense form.

Hence was it we found imagination so important an element in the recollective process.

2. These ideas or sense forms are distinguished from sensations by less intensity. Only in cases of hallucination do they have equal or greater intensity. Gaze at a friend and study well her countenance; now close your eyes, allow the after-images to fade, and then, still with eyes closed, visualize her face: you will at once learn how inferior to perception is the idea or sense form. The English painter who could call up images of his sitters, even when they had been before him for only half an hour, so that he could perfect their portraits in their absence, was mentally unbalanced, soon lost power of distinguishing imaginary from real persons, and spent thirty years in a madhouse. Sense forms in normal minds are weaker than sensations.

3. Ideas or sense forms originate in the sensori-motor ganglia. Destroy these, and imagination ceases as certainly as sensation and perception. Extirpate the optic ganglia, and you not only fail to see, you fail to imagine anything as seen; you can not even imagine darkness.

Persons of unusual facility in visualizing may by intensely thinking red cause a complementary green on the retina of the closed eye. It is hard to imagine a labial with the lips apart. The author, in recalling the very distressing incident of seeing one of his sons sink into the water of a swimming pool beyond his depth, always finds himself tending to draw up his own body just as he saw his son do it; he remembers the incident not only in visual, but also in motile sense forms.

This fact furnishes a method of classifying imaginations. They are visual, audile, tactual, motile, according to the kind of sense forms mostly used. Persons extraordinary in one of these are very likely to be deficient in the others.

4. Hence the realm of imagination is the whole universe of the created, the limited, the composite. It does not and can not include the Deity which is uncreated, unlimited, incomposite—the sense forms can not picture God. All pictures of the Deity are anthropomorphic, and so mere approximations and distortions. We can think and know God, but not by imaginative processes.

This is the truth in modern agnosticism. The very fact that agnostics argue about an Infinite proves that such a being is thinkable and tentatively knowable. We could not predicate ignorance of something which could not come into thought.

5. The material of the imagination being the endless variety of the universe, to equip a mind for its best ideation, observation must be lively and the store of knowledge full. An ignorant poet—an Ossian or Homeric druid—may have an intensity of ideas, but these are simple, few, and oft-repeated. A sublime spirit, high as heaven, wide as the horizons and deep as ocean, requires as a feeder the keen observation of a Shakespeare, or the learning of a Milton, a Coleridge, or a Goethe. Well said a French wit, "The soul of the poet is the mirror of the world."

6. Since the mind can construct and create in sense forms as well as restore, we usually speak of the reproductive, the constructive and the creative imagination.

The reproductive serves recollection.

The constructive rearranges, readjusts, divides and joins together.

The creative discerns the as yet unthought.

Or we may classify imagination according to the method of its processes, and declare it natural, logical or poetical. In the first case it follows Nature's order of suggestion and association; in the second, the logical sequence, working inductively or deductively; and in the third it aims at poetic effect by appealing to the sense of the beautiful.

But always the ideation will reproduce, construct or create, the selection of methods of proceeding being largely a matter of disposition, education, etc.

7. Figures of speech are very common instances of the play of the constructive function. "Intelligence rarely allows itself in speech without metaphor; we seldom declare what a thing is, except by saying it is something else." (George Eliot.) As when Ulysses compared Nausicaa to a young palm tree springing up by the altar of Apollo, as when Wieland declared that his soul was as full of Goethe as the dewdrop of the morning sun. Names of things are almost always imaginative, as when the Coreans call flame the fire flower, and say, when they want you to strike a light, " Make the fire flower blossom." The ancient Mexicans and Peruvians beheld gorgeous humming birds glittering over gay flowers—motion on wings—symmetry radiant—gems aflight—flashing emerald, ruby and sap-

13

phire light, and they called them names meaning "rays
of the sun," and "tissues of the day-star," "living sunshine," and "day-star light." Language is full of
such poetry. Of course much more is involved here
than the mere conception of the sense form; but it is
the constructive ideation in the general process, which
leaves on the words their unfailing phosphorescence.

8. Dreaming is the play of constructive imagination, enhanced in its exuberance by the very fact of
the withdrawal of the primary control. Ideas become
both relatively and absolutely more intense than in
waking moments.

Somnambulism, natural and induced, is but a
dreaming vivid to action, in which imagination is
quickened and controlled from without.

Hallucination finds this constructive function its
very organ. Reverie is a kind of self-induced hallucination, in which we dream while awake. Ordinarily it
is passive, involuntary in its play of fancy, and enfeebling of intelligence.

The constructive imagination of the brain worker
is very different from this delicious *träumerei*. Reverie is a train of fancies following the line of least resistance; literary or artistic or musical composition
seeks lines of greatest resistance, and is work involving severest mental discipline and resulting in substantial mental products.

So it goes,

"The lunatic, the lover, and the poet
Are of imagination all compact.
            The lover frantic
Sees Helen's beauty in a brow of Egypt:
The poet's eye, in a fine phrensy rolling,

Doth glance from heaven to earth, from earth to heaven,
And in imagination bodies forth
The forms of things unknown : the poet's pen
Turns them to shapes and gives to airy nothings
A local habitation and a name."

9. Imaginative creation is not true of the human mind, except in a relative way. It is simply the creation of sense forms never conceived before, and depends not so much upon vivid power of imagination as upon splendid gifts of elaborative genius. Of this genius we will speak in another connection.

10. The intensity of ideas depends not upon the intensity of the occasioning or original sensations, but upon the amount of feeling originally stirred by those sensations. The idea bright is not necessarily more intense than the idea dark. Those ideas are vivid whose accompanying sensations enlist our interests to a marked degree; and such are most likely to recur in recollection and in construction.

Intensity may reach that dangerous pitch where sense forms are mistaken for things perceived; then the superstitious see ghosts and hear mysterious noises, cowards start at their own shadows, murderers are haunted by visions of their pale victims.

Pessimism enlarges evil and minimizes good, while optimism enlarges good and minimizes evil. The imagination is a telescope of all powers, and you may use either end.

Many of the fond delusions of mankind concerning its own destiny are so accounted for. The dream of a golden age—far back in the past—and the expectation of a golden age yet to come—far forward in the future —only emphasize the innate imaginativeness of that

universal humanity, in which faith never dies, in which
"hope springs eternal." " He who has imagination
without learning, has wings without feet." (Joubert.)

CHAPTER XXXVII.

THE COMPARATIVE PROCESSES—CONCEPTION, JUDG-MENT, AND REASONING.

1. WE combine these three processes, because they
all give exercise to the one psychic power of comparison. Conception compares ideas to form concepts; judgment compares concepts to form proposi-
tions; reasoning compares judgments to reach conclusions. The difference between a judgment and a conclusion has been thus stated: " A judgment is knowledge that is reached, a conclusion knowledge that
becomes."

2. Conception. The mind compares ideas of a
kind, and abstracting qualities, possessed in common by
all, forms a general notion or concept. This process
seems to depend upon the persistency of mental states
and their natural affinity for one another. It pertains
to the grouping function of consciousness. Ideas of a
kind blend together and become, to use Romanes'
figure, a sort of composite photograph in which in-
frequent characteristics fade into indistinctness and
common features alone appear. Thus the concept
" tree " is the result of long observation of trees and
the constant occurrence of the ideas of this maple and
that beech and yonder pine. A composite idea, which

is all trees in general and none in particular—in short, the general notion "tree"—has thus arisen. Psychologically speaking, such words are concepts; in common parlance they are terms.

3. To concepts may be attributed quantity, quality and relation.

In quantity, they have extension (area, *Umfang*) and intension (content, *Inhalt*). The greater their extension so much more numerous the individual notions they represent; the greater their intension the more varied their qualities and restricted their range. "Man" is more extensive than "orator," because orators are only a small class of men; but "orator" is more intensive than "man," as the orators have all the qualities of men besides those of their own class.

In quality concepts are clear or obscure, distinct and indistinct. They are clear when we can well discriminate them from other concepts. They are distinct when their individual parts can be discriminated one from another.

In their relation to one another they may be classed in matter of extension as exclusive, coextensive, subordinate, co-ordinate, or intersecting; and in matter of intension as identical or different.

4. Judgment compares ideas, perceives a relation, and forms a proposition. A judgment is thus the expression of a perceived relation; it, too, like the concept, has quantity, quality and relation.

In quantity, judgments are universal—"All men are mortal"; or particular—"Some men are virtuous"; or individual—"Cæsar was a conqueror."

In quality, they are affirmative or negative.

In relation, they are categorical or conditional,

and, if conditional, then hypothetical, disjunctive, or dilemmatic.

5. Reasoning is simply a comparison of judgments, with inference—as Romanes defines it, " the faculty of deducing inferences from a perceived equivalence of relations."

For the various methods of doing this, wisely and unwisely, we must refer the reader to the manuals on logic.

6. To say that man is the reasoning animal does not properly differentiate him from the brutes. All vertebrates form judgments, and probably they combine them into reasoning processes. We believe that in dogs, pigs and horses intelligence attains a high degree of ratiocination. Insects also are quite rational, the ants in their logic seemingly surpassing many races of savages.

7. Inferences which are complex to a feeble mind may be very simple to one larger and keener. It takes a long while to convince a savage, used to counting no more than five on the fingers of one hand, that six times five are thirty. A civilized child, however, perceives this as a self-evident truth, and, without process of calculation, asserts it as axiomatic—as simple to him as the proposition " The sky is blue."

It seems likely, then, that all processes of inference are but the crutches of a halting mentality. And thus reason, instead of differentiating man from the brute, rather shows his inferiority to possible higher intelligences. It is not extravagance to surmise that what seems to us obscure conclusion of long and tedious reasoning may oftentimes be of a simplicity to appear luminous to a higher grade of mind. The child spells its one-syllable word letter by letter; the scholar, however, reads by sentences and by paragraphs. The savage counts five on the fingers of one hand; a Zerah Colburn instantly on demand gives the correct results of complicated stupendous mathematical computations. Just so, while the uncultivated thinker works out his conclusions by tedious process of syllogism, genius sees the result at the beginning, and the highest processes of logic to the man of to-day are but the a b c of the man that shall be, and the clearly seen axioms of mind suprahuman.

8. Just here genius asserts its superiority. It has been very generally associated with the creative imagination, as without a high degree of the latter its marvelous outbursts would be clearly impossible; but in essence genius is an extraordinary assertion of judgment. It sees with intensity its images, but much more it discerns inferences with prevision. Says Ruskin : " Hundreds of men can talk for one who can think, but thousands can think for one who can see. To see clearly is poetry, prophecy and religion all in one." Genius looks in many directions; it may be poetical, literary, musical, artistic, political, military and metaphysical. It has been noticed of valuable discoveries that they are so simple, and we wonder that even we ourselves did not perceive the obvious fact. They awaited an eye that could penetrate their secret. Watt's steam engine, which seemed an intuition, was only the inference of genius. All great scientific discoveries are made by the leaping inference of genius before the actual demonstration. Columbus discovered America before he left Palos; Leverrier saw the planet Uranus long before he gazed upon it

through the tube. Napoleon was a great general, because he could readily foresee every possible combination of his enemies, and every possible contingency; the batteries of his unerring judgment won the battle ere his brazen artillery opened fire. Bismarck and Gladstone are great in genius, because they judge more simply and more accurately than other men. Back of all immortal verse, there is not only the beauty of art, but also the sublimity of lofty inference.

The creations of genius may seem flashes out of absolute darkness, epochs in the history of thought, but in reality they are only the discoveries by great minds of facts Nature is hiding. The poet, painter, dramatist, or prophet detects in the ordinary and transient its elements of ideal and imperishable beauty, truth, or goodness, and constructs a poem, a painting, a theory, an invention, a parable in righteousness, that needs must live because a revelation of the universe to human nature or of human nature to itself.

9. Genius finds its response in minds of a second grade that can not be called " creative." Geniuses are few and far between; their lives are a " sublime storm "; they are isolated like lighthouses from the rest of the world, lifted up like snow-capped mountain peaks into the heavens and into the gales. More numerous are those lesser natures that understand, interpret, and teach the world to adore them. Let the Deity give to whom he will his signet ring, but let no gentle or noble nature fail to aspire to this second glory—to understand, to interpret and to defend.

10. Wit is a keen play of the comparative processes discerning apt but unusual relations and contrasts be-

tween things. Its simplest and easiest form is the pun
or play upon the varying meaning of a word, which may
be uttered in soberness to emphasize a truth, or in mer-
riment to point a joke. In the Hebrew Scriptures the
use is serious, as when Samson punned upon the jaw-
bone in his song of triumph, or when Jesus compared
the Spirit to the wind. No better punning of the second class can be cited than that of Senator Evarts,
who, when Lord Coleridge demurred at the old story
of Washington having in his youth thrown a dollar
over the Rappahannock, replied : " But, sir, a dollar
would go farther in those days than now ; and, moreover, it seems less improbable that he threw a dollar
across a river, when you reflect that he threw a sov-
ereign across the sea ! "

Proverbs, epigrams, aphorisms, and all short, pithy
sentences, are but witty, humorous, or satirical judgments. Alger declares aphorisms " portable wisdom,
the quintessential extracts of thought and feeling."
An old Latin recipe for a proverb prescribes that it
must be " like a bee, short and sweet, and with a sting
in its tail."

Wit often marks great minds, and generally it
accompanies genius ; sometimes, however, its keen
incongruity is rather an evidence of oddity or mental
unbalance—as in the case of the lunatic who said
to a visitor in the asylum, " Sir, I am Alexander the
Great " ; and on a subsequent occasion, to the same
gentleman, " Sir, I am Napoleon Bonaparte." " Oh !
but," expostulated the visitor, " the last time I was
here you were Alexander." The lunatic mused a moment, tapped his head thoughtfully and responded,
" Did I ? Well—that was by my first wife ! " Or as

in the case of that Irishman who, when accused of cowardice for running off the field of battle, said, " Shure, and it is better to be a coward for five minutes, than to be dead intoirely all the rest of me life !"

It has been claimed that wit must be spontaneous, but this does not at all follow. Professional wits find it necessary to resort to premeditation. Dean Swift would lie abed wide awake, mornings, preparing extempore flashes of wit for the day. Washington Irving would swing on gates on sunny mornings, busied in the same kind of industry.

11. Humor adds the element of absurdity, which is as truly interwoven into the very fabric of the universe, and into the warp and woof of history, as the beautiful, the sublime, or the painful. It is a habit of viewing things on their grotesque side, and is separated from contempt by a certain element of sympathy. It dwells upon the whimsical and fanciful in life and character. While wit does not by any means always cause laughter, humor never fails to provoke, if not laughter outright, at least a smile. The infinite humorsomeness of Sheridan appeared when, the night his theatre burned to the ground, he quietly seated himself in a public house near by and in sight of the flames sipped wine calmly. Expostulated with by friends for this astounding indifference to the wreck of his fortunes, he calmly replied, " Surely a man may be permitted to drink a glass of wine at his own fireside !"

12. Satire is judgment reflecting the irony of fate, the inconsistent in character and pitiful in destiny forming its subject; for mockery seems omnipresent

in Nature and in history, and the satirist only echoes the bitter laugh of a somewhat in the universe itself.

The Turks say, " Cross the sea and drown in a brook." The Chinese assert that " Order the coffin and the man won't die." And the Spaniards reflect that " The worm has ever a poor case when the chicken is judge." A Saul falling on his own sword in Mount Gilboa just as Israel is about to obtain glory, a David fleeing from his own son Absalom, a Marius amid the ruins of Carthage, a Napoleon at St. Helena, a Christ crowned with thorns and enthroned upon a cross—in such themes does satire delight.

13. Comparison is also the faculty of classification, which is only an enlargement of the process of conception. Discerning likenesses between things, which therefore seem allied, we group them together in a class and name them with a word. Discerning likenesses between these classes, which therefore seem allied, we again group these together in a higher or more comprehensive division, and name this with a word. So we proceed, our describing name ever becoming more extensive and less intensive. Men, women and children—black, red, brown, and white —become " man "; then man, monkey, ape, etc., become the simian ; then the simian, the ruminant, the carnivore, etc., become the mammal ; then the mammal, the reptile, etc., become the vertebrate ; then the vertebrate, the mollusk, the annulose, etc., become the animal. The individual notions, Tom, Dick, and Harry, cover little ground but many qualities. The common notions, man, monkey, ape, etc., lose depth but gain in amplitude. Classification, as it proceeds,

spreads out the notion until it becomes quite diaphanous.

This is the method the mind pursues to enable it to grasp a multitude of objects; it differs from mere enumeration, in that the latter simply grasps at numbers. We comprehend numbers in groups, ten units one ten, ten tens one hundred, etc.; but the groups indicate nothing more than collective enumeration. Classification, pursuing the same methods, uses a descriptive nomenclature and groups on system.

Again we are reminded of the limitations of our knowledge. As mind must reason step by step, so it must hold and master its knowledge group by group. Hence the necessity of libraries, with alcoves and catalogues; of museums, with rooms, cases and descriptive labels; and, in the mind itself, of pigeon-holes and a mental index. We may call classification the method of learning—the system of knowledge.

14. A word on the communication of judgments. Had men been created, or indefinitely continued to be, solitary individuals, mating only for a season, as do most animals, a hundred times the period of human evolution would not have sufficed to impart civilization. Man needed not only the heritage of ancestral judgment, but the assistance of social common sense. As a social being he learns to express judgments readily, and in the interchange "mind sharpeneth mind."

It is significant in this connection that social animals have at least rude methods of exchanging judgments, and their progress toward civilization is always in exact ratio to their powers of expression. Ants have a ready (though unknown to us) method of thought-transference, and hence a high degree of civilization;

and even buffaloes, wolves and marmots understand each other.

T. W. Cowan, a close and accurate observer of bees, declares that the voice organs of his pets are three-fold—the vibrating wings, the vibrating rings of the abdomen, and a true vocal apparatus in the breathing aperture of the spiracle, the first two producing the buzz, and the last the hum. He believes that he has truly interpreted the various significant sounds. Hum-m-m is the cry of astonishment. Wuh-wuh-wuh glorifies the incessant accouchements of the queen. Shu-u-u is the frolic note of young bees at play. S-s-s-s means the muster of a swarm. B-r-r-r is the appropriate call in slaughter or expulsion of the drones. The tu-tu-tu· of the newly hatched queens is answered by the qua-qua-qua of the queens still imprisoned in their cells.

Similar discoveries, with help of the phonograph are claimed on behalf of the monkeys by Garnier, and the facts proved certainly deserve closest attention.

15. The history of expression began with the glance, the growl, the purr, the wagging of the tail, the gesture. Later came facial movement, chiefly teeth-gnashing, knitting of the brow and grimaces. In man, this speech of bodily signs attains its greatest perfection in the infinite variety of eye-glances, of feature-movements, ingenious shruggings and significant gestures. All sign language can be grouped conveniently under three principles:

(1) Of altered innervation, as when strong emotions react upon central organs so as to cause tremblings, blushings, erections of hair, etc.

(2) Of analogous sensations, as when we express disgust by posturing the mouth and face.

(3) Of significant motions, as when we wrinkle the forehead to suggest a wrinkle in the mind, or point heavenward to bring to mind an overruling Providence (Wundt).

16. Verbal language is the flowering of expression; and once invented it becomes the prime cause and the most accurate measure of human evolution. "A language is to be considered the collective brain of a nation: the vocabulary shows the richness of its ideas, the syntax how it works them."

In its beginnings verbal language is strongly physical, like sign language; for men learn to speak long before the age of self-study. "All roots are expressive of sensuous impressions only; and all words, even the most abstract and sublime, are derived from roots" (Müller). These roots indicating activities became subject, verb and object by variations in sound or diacritical additions. As thought gained in powers of discrimination, adjectives and adverbs were introduced to name perceived qualities, and to indicate the how, when and where. Prepositions came to express relations now more and more manifest. Finally, conjunctions indicated the working into language of the laws of thought. Elaborate syntax and affectation of style must have been, in all cases, the result of high civilization.

Hence, to study the language of a people, from its rude beginnings, is to unearth their history, their customs, their social, political and mental growth. The comparative study of the Aryan languages has given scholars a very fair acquaintance with the conditions

of an ancient people, whose story was long since lost,
even to tradition. A close scrutiny of the word-forms
and syntax of any race will restore their logical processes and rhetorical conceptions. Indeed, the history
of philosophy will never rest content with records and
legends; it will go far back of all remains to the revelation of word derivations and linguistic forms in
primitive languages.

Hence, in general, the great value of linguistic
studies. They teach one not so much to speak as
to think. Our own processes of judgment are thus
multiplied, varied and corrected, by comparison with
the mental processes of other peoples and ages; while
the comparison itself, with its necessary research and
nice balancing of many considerations, is a discipline
in both analytic and synthetic thinking of supreme
value.

17. The expression of thought assists the formation of thought: so that some thinkers have boldly
claimed that there is and can be no thought without
language, and no growth of thought without expression—notably Max Müller. Ribot goes so far as to
assert that thought is a word or an act in a nascent
state. However this may be, if you would disentangle
a confused mass of reasonings, put them down in syllo-
gisms. If you would think clearly on any subject,
talk it over and force yourself to write out your
thoughts. The labor of expression will be found an
effort of the mind to analyze and arrange its own processes. Painstaking composition greatly expands and
simplifies intellectual operations.

18. We have found all the other psychic powers
imperfect in working; the comparative processes are

no exception in this regard. They may work delusively, under sway of prejudice, superstition, skepticism, etc., and inferences as well as judgments may be ill-formed, unbased in logical verity, and if not whole lies at least only half truths. The history of fallacies, in simple and compound judgments, would be almost the whole story of human fraud, delusion, folly, prejudice, superstition and cruelty.

CHAPTER XXXVIII.

FORMAL THOUGHT.

1. KNOWLEDGE is of two kinds, empirical and formal. The former is special, incidental, phenomenal; the latter is universal and necessary.

Formal knowledge comprises the laws of thought, feeling, and volition, or, in other words, the limitations and inner necessities of psychosis.

The Greeks named this mental realm the νοῦς; by the Germans it is styled the Pure Reason; while to the Scotch metaphysicians it has been the Regulative Faculty.

2. The limitations and necessities alluded to are purely subjective. Psychic action presupposes them. No amount of perception or length of experience could supply them, even by heritage. They are not necessary because the mind so regards them; but the mind so regards them because they are necessary.

3. The senses supply the material, the reason supplies the form of thought. Without the senses and the sense habitudes, mind would be utterly vacant:

without reason, mind would be mere chaos. First, when reason regulates sense, have we cosmos.

4. Formal thought is the last object of contemplation in self-study. Animals, savages and children of this think never. Such thinking as they do, of course, conforms to the laws of mind, but they do not reflect upon this fact, nor classify nor even recognize these principles. Hence, though these ultimate data are necessary—*semper, ubique et ab omnibus*—yet are they by no means ever, everywhere and by all perceived.

Descartes argued thus, human existence: *cogito ergo sum*, "I think, therefore I am"; and it is true that thinking is the best logical condition of faith in one's own being. But, as a matter of fact, man says *sum*—"I am"—before he declares *cogito*—"I think"; and it is last of all he asks himself how and why he thinks. The study of the inner limitations and necessities of psychosis is the last and the highest and most difficult effort of the mind.

5. There are three ultimate forms of consciousness—knowing, feeling and willing. Attempts have been made to reduce these seeming ultimates to unity but in vain; the result has proved supersubtle, improbable and utterly barren.

6. Each of this trinity in action runs for its own goal.

Knowledge compels us to the true or the false.

Feeling has in further view the beautiful and the foul.

Willing ever involves some aim of good or bad.

The true, the beautiful and the good are as ultimate seemingly as knowing, feeling and willing. Here again efforts have been made to secure unity,

14

but only by a sort of metaphysical violence which is as unjustifiable as it is valueless.

7. Embracing all knowing, feeling and willing are the two ultimates, Time and Space. Here again attempts have been made at derivation. Thus Wundt claims that space is a synthesis of local signs and movement. Duration and extension you may define and analyze, but not time and space. If you had not these ideas already in stock, you could not by any amount of experience evolve them; indeed, did they fail, you would do no thinking at all. What has been shown has been not the genesis of the formal thought, but only its slow dawning upon human consciousness as thought.

8. The laws of thought as thought—that is, the laws of logic—are also regulative functions of the mind. They could not have been evolved, for they are presupposed in thought itself.

(1) Of identity. $A = A$.

(2) Of contradiction. Of two contradictories only one can be true. $A - A = 0$.

(3) Of excluded middle; which compels us, of two contradictories that can not both exist, to think of the one or of the other as existing. A either is or is not.

(4) Of reason and consequent, or there is reason for every inference. A is, because. This is the law of logical cause and effect, and at least analogous to the physical law of the same title.

9. That particular things are or are not beautiful, formal thought does not inform us; for what is pleasant to one may be disagreeable to another, and what is foul to this man may seem fair to that; but that the beau-

tiful is beautiful and that the foul is foul is regulative necessity. The laws of taste and sentiment, though as yet feebly discerned, are at least emerging into conscious importance, and beyond reasonable question are not conventional, but as inevitable and imperative as the laws of logic.

Wundt claims that beauty may be reduced to the idea of order. The beautiful is the orderly. Likewise the moral is simply that which is useful, and religious sense is a mere process of reasoning or analogy. All such vain efforts .mistake the history of the dawning of sentiment, morality and religion upon the human mind in savagery, for the basic principles of sentiment, morality and religion themselves.

10. The laws of volition are likewise ultimate and in the very nature of choice.

That man is free to choose (in narrow limits), and that responsibility accompanies such choice, in right doing and in wrong doing, is clearly basic fact, forming a " moral consciousness," a realm of conscience.

11. As the mind progresses in self-study, it is clearly seen that three ontological facts loom up in the background of all thinking—being, infinity, eternity; and these three coalesce in a consciousness of Deity. All highest philosophy faces this sublimity, and though in crass and ignorant minds it be latent, the truth even here asserts itself in a yearning toward the infinite and in a universal religiousness. For this Being is not living matter, and this Infinity is not indefinite extension, and this Eternal is not endless duration. Living matter is limited, being is absolute. Indefinite extension is made up of parts, endless duration is composed of successive moments. Eternity is incom-

posite, outside of past, present, and future, a unit. Infinity is not composed of spaces and distances, but undivided and indivisible, an absolute Unity.

You can not prove the existence of a Deity by any reasoning process, for there may be nothing in a logical conclusion, which was not in the premises; and if God be in your premises, you have begged the question. If he be not in your premises, he will not be logically found in your conclusion.

Man is the only animal who has attained sufficient psychic expansion for this recognition of the Divine; and of men, only few have risen into clear discernment of the Absolute; though probably none are so low as not to respond to some influence of this presence.

12. There are certain objective necessities, learned by experience, that have come to sway thought in a formal way, much as if originally subjective. Thus the distinctions of matter, life and mind; also mathematical and physical laws, which at first observed acquire the force of postulates. "Two and two make four." "Two particles of matter can not occupy the same space at the same time," etc.

In this class of intuitions probably belong personal identity and personal unity. That I am myself at all times and in all places, and that I am one person— these seemingly necessary postulates are no doubt results of long human experience and reflection—facts discovered by the race, but born as necessary forms of thought in the individual.

CHAPTER XXXIX.

REVIEW.

1. STRICTLY speaking, there are no faculties, and only the mind in differing phases of psychic action. What we name such are simply constitutional modes of mental behavior.

2. Every mental act or state is complicated with other mental acts or states. We analyze psychic phenomena and describe the simple elements, but the phenomena themselves are never simple.

3. Notwithstanding this complexity, the searchlight of consciousness can not play upon many groups of phenomena at once. If any but confused thinking is to be done, attention must be directed to one grouping, and abstraction for the time from all others enforced. The success of Hegel is in part explained by the fact that he took a manuscript to his publisher in Jena on the very day when the battle of that name was fought, and to his amazement—for he had heard or seen nothing—he found French veterans, the victorious soldiers of Napoleon, in the streets. Mohammed falling into lone trances on the mountains above Mecca, Paul in Arabia, Dante in the woods of Fonte Avellana, and Bunyan in prison, form eloquent illustrations of the necessity of mental seclusion and concentration in order to arrive at great mental results.

4. The relative vividness of mental states has much to do with defining their locus. Perceptions are in general more vivid than memories, and both in

general much more vivid than concepts. Ordinarily attention will prevent hallucination.

5. Psychic acts or states often perish stillborn or later in infancy, or still later in childhood or youth. Many impressions on end organs do not stimulate them sufficiently to secure a message to the brain; hence no resulting sensations. And, then, many sensations are not vivid enough even to enter the subconsciousness; hence no resulting perceptions. Many perceptions are too faint to link themselves to common trains of thought; hence, pushed aside, they are unused by imagination or comparison. Many ideas are not intense enough to attract attention; hence no judgment based on them. Many judgments are exceedingly feeble or inaccurate; hence no inferences.

6. How can even the most cautious thinker assure himself of accurate acquaintance with his surroundings in view of the proved imperfection of all our mental apparatus? He can not! All sciences have to be regularly readjusted every few years. We are none adepts, and all novices. As Socrates said in Phædo, "Many are the wand-bearers and few are the mystics."

7. While there is an evolutionary history of the cognitive powers, there is no evolution of the powers themselves. They all seem at least nascent in the early forms of life. The gradual development of the faculties connects itself with the history of the growth of the end organs and of the central nerve masses, and with the fact, never to be forgotten, of the withdrawal of consciousness from lower to higher nerve ganglia.

8. The evolution of a psychic function ought to cast no doubt upon the knowledge it has in time com-

municated. Seven and six make thirteen, even if it be proved that our ancestors, like many savages of to-day, could not count more than five on the fingers of one hand. The small beginnings of science, art, ethics and religion do not in the least discredit the true, the beautiful and the good. Psychic evolution has been a finding out. The Parthenon is no less beautiful as a temple, nor the Principia profound as a book, nor the death of Joan of Arc sublime as heroism, because the ancestors of architect, mathematician and enthusiast were clothed in skins, and ate raw flesh, living in holes, counting fingers, and burying alive their aged parents. Plato was once an unconscious babe; but he had become a man of sublime intelligence when he wrote the Phædrus and Symposium. Shakespeare was once an infinitesimal droplet of protoplasm, but he came to be the supreme genius of literature. Just so the human race had its babyhood, and knowledge confesses to a slow development, but the resulting wisdom is not therefore vain.

THE FEELINGS AND THE WILL.

CHAPTER XL.

THE FEELINGS.

1. FEELING is a primary mode of psychic activity, and, being ultimate, admits of no definition, and much less of analysis. Feeling is to feel! It is not to be confused with sensation, which, though usually toned by the pleasurable and the painful, is in itself entirely intellectual.

2. Feeling has had its evolutionary history. It appears simply in the lowest forms, and becomes complicated in the higher only with multiplication of nerve masses and general psychic development. As knowing begins in naked protoplasm, a mere sensitiveness to excitation from without, and willing in a mere self-contractility, so feeling commences in protoplasmic experience of pleasure and pain. The infusoriæ seem to suffer and to enjoy, to love and to hate, to hunger and to be angry. Insects prefer the beautiful, and birds appreciate fine music and gay plumage. Love of young and of mate is strong even among very humble creatures. Sense of right and wrong, vanity, pride, self-righteous-

ness, hypocrisy and remorse are common among mammals.

3. Feelings are characterized by tone, strength, rhythm and content. These will be considered in order, in the following sections.

4. By tone we mean pleasurableness or painfulness; that is, the mind is never in a condition of indifference emotionally toward any of its mental states, and its emotional interest always involves more or less of satisfaction or dissatisfaction. Why this is so we can not say; it is an ultimate fact.

(1) The zero point between pleasure and pain is variable with time and clime, with individuality, and with bodily and mental condition. A joy for one is a pain for another, a pleasure to-day will be sorrow to-morrow, a sport in winter may be torture in summer.

(2) Tone depends upon the relative intensity and upon the quality of the mental states in question. Experiences ordinarily delightful, if too intense, give rise to pain.

(3) The conditioning quality depends upon a variety of laws. Pleasure may result from the gratifying of a mere physical craving, as of hunger; or it may indicate harmony with æsthetic or moral impulses. Emotional quality may be high or low, good or bad, and can be understood in its workings only when we can trace it out with help of close study of past history, individuality and environment.

5. Feelings themselves are characterized by relative intensity, which is conditioned by the vigor of exciting causes, the general vitality, the current healthfulness, etc.

6. Feelings are said to be rhythmic, because subject to periodicity. They rise and fall with the rhythmic movement of the nerve cells they use. As feeling makes severer demands upon the nerve reservoirs than cognition, exhaustion follows more quickly: an explosion is speedily succeeded by depression. Rapid and high elevation above the zero point is followed by rhythmic fall below that point. Children who laugh before breakfast cry before night. An ecstasy of joy reverts into a nausea of satiety or an agony of melancholy. The feelings of tropical people are violent, because, usually leading a sluggish life, their batteries of nerve force become heavily charged, and so capable of tremendous explosion.

This rhythmic peculiarity of feeling explains the superficiality of overemotional natures. Reaction follows soon. When the three comforters came to Job to convince him that it served him right to suffer woe, Bildad, the fiercest, was the soonest silenced.

Explosions often relieve nervous tension caused by pent-up force, uneasy to be released. To weep profusely relieves sorrow; to grumble solaces discontent; to slam a door gives vent to wrath; to pray helps the penitent to peace; to sing and leap and laugh afford relief to overjoy.

Herbert Spencer has called attention to the fact that expressions of emotion by dancing, poetry, or music always have assumed a rhythmical character.

7. The content of feeling is as hard to classify as the mind's treasury of thoughts. The variety of feelings is as endless as of thinking. Various classifications have been attempted, but with questionable success; about the best we can do is to describe them as

sensuous, æsthetic, intellectual, or moral. The dividing lines, however, are indistinct, and many feelings can be named that belong to two or more of these classes. Thus the emotion excited by a delicious strain of music may be intellectual, ought to be æsthetic and surely is sensuous.

8. Neither the tone, strength, rhythm nor content of any given feeling on any given occasion is absolute, but conditioned on circumstances, individuality, temperament and point of view.

9. Feelings may be conscious or subconscious, automatic, reflex, or voluntary. In other words, they interpenetrate all psychic activities, at all times and under all conditions. Waking and sleeping, in higher and lower nerve centers, and in every kind of nerve utterance we feel.

10. Feelings furnish coloring and tone for intellection. Even if the content of a thought be purely intellectual—and it seldom is—the motives and aims are sure not to be so. Sensuousness, prejudice, and æsthetic and moral considerations interpenetrate everywhere; hence the danger of inaccuracy in the working of all the cognitive powers. Facts are distorted and judgments shaded and arguments vitiated by feeling. And then, on the contrary, the pursuit of truth is stimulated, fraud is abhorred, lies avoided, and facts asserted courageously, because of feeling. It curses and it blesses.

On the whole, feeling is as safe a guide to reality as thinking. The beautiful and the good are as ultimate as the true, and feeling is quite as likely to be normal as willing or thought. To deny the objectivity of beauty and right and the validity of taste and conscience is mere skepticism, leading only to psychological absurdities.

Feeling needs education, and receives it in artistic study and moral and religious restraints and exercises.

11. Feelings furnish motives for action. We eat because we are hungry and fight because we hate, sigh because sad and rave because in love. Without such prompting there could and would be no action at all. A pain, a want, a desire, an antipathy, an affection, an enthusiasm, must antedate all voluntary or involuntary, conscious or subconscious, action.

12. Feelings have their language, and when thus expressive are called emotions. They utter themselves directly in interjections, and indirectly in other forms of speech through accent and emphasis. Facial expression, gesture, and the well-known outlets for passion, all convey to others easily our liking, love, hate, scorn, admiration, sympathy, or delight in truth, in art and in righteousness. Uncultivated natures are very demonstrative because they fail of education, which, teaching self-control, suppresses any frank display of feelings. Culture, developing the language of thought, represses the language of feeling.

13. The diseases of feeling belong to three classes: incongruity, excess and lacking—the wrong kind, too much, and too little. Herein find explanation most of the unnecessary agonies of men, and much of the vice, crime and pain with which they torment themselves and one another.

CHAPTER XLI.

WILLING.

1. THE will is a name for the self-determining function. Like knowing and feeling, it is a primary mode of psychic activity, admitting neither of analysis nor of definition. Willing is to will.

2. It appears in mere naked protoplasm as a self-determined contractility. In zoöspores, spermatozoids, etc., it attains a variety of action. In animal and vegetal persons it occurs as a common function, controlling the general movements of the protoplasms in contact. With the appearance of nerve cells and muscles, its range both of excitation and of execution is vastly enlarged. Indecision, resolution and willfulness are to be found in all the higher animals.

3. The end organs of willing are the muscles, and the media of control are the efferent or motor nerves. This machinery is worked by discharge of force generated in nerve cells, the act of discharge being consciously or subconsciously volitional. Such force is derived from the break up of the highly complex and unstable nutritive material supplied by the blood, and is the release of a kind of explosion. (See p. 27.)

Hence vigorous self-determination depends upon plentiful and wholesome blood supply, or ultimately upon good food well digested and good air well inhaled. The secret of energy, and even of ethics, in the last analysis, is largely in sound digestion and good

ventilation. Lessen or vitiate the supply of blood, and
you may produce any desired degree of inaction and
helplessness. On the contrary, cerebral congestion in
a vigorous person (as in the insane) may generate tremendous outbursts of muscular activity and stern resolution.

4. Willing, in intensity, ranges up and down a
scale in which are three degrees—wishing, purposing
and determining. Weak volition wishes, resolute volition purposes, while strong volition acts.

5. Willing may be automatic, reflex, or voluntary,
and may take place in single cells or in groups of cells,
consciously or subconsciously.

6. The diseases of will are: (1) Indecision, culminating in inaction, dreams, reverie, hallucination, hypnosis; or (2) willfulness, culminating in insanity; or
(3) perversity, culminating in self-determination coun-
ter to higher motives and in defiance of laws of rec-
titude.

7. Willing is powerfully controlled through the
feelings by the cognitive powers. An intelligent
person acts in harmony with his mental judgments,
an upright one in obedience to his moral judgments, a narrow mind in servitude to his prejudices, and a fool in chase of his dreams, whims and
fancies.

On the contrary, willing exercises a powerful control over both the thinking and the feeling. It can
regulate thought by calling upon the appropriate faculties and forcing them to do their work; it can in-
flame or restrain passion by discriminate manipulation
of the proper nerve centers. Were this not so, no mental work would ever be done and no moral accounta-

bility ever incurred. We can play upon our cognitive and emotional natures much as a musician upon his instrument, the while he—and we—are ourselves affected by our own music.

8. There is no willing without motives furnished by feeling. The most absorbing and important controversy in all ages has been whether these motives control, necessarily, as absolute causes of action, or whether mind has any real power of self-determination. It is admitted by all that motives occasion; but do they determine? The automatists, and all materialists, Stoics and fatalists, and among theologians the Calvinists who believe in absolute Divine decrees, say yes. But consciousness, conscience and common sense, denying this atrocity, affirm human freedom. The range of choice may be very narrow, but within this range the will is free. The method of this freedom is a very great mystery. It seems to involve an act of causation, a new beginning, in each determination. It introduces into human affairs an element of caprice, which, if the sphere of choice were less limited, might prove fatal to consistency and remove history from the domain of science. Yet, undoubtedly, that element of caprice is present in all human conduct, and, as Froude the historian admits, history is not and can not be an exact science.

9. As the grade of being rises, self-determination becomes less limited by conditions. Among protozoans, freedom and necessity must be almost interchangeable terms. The same is true of our own individual cells and lower nerve centers. As personality becomes emphatic in the progress of evolution, the range of freedom enlarges. With this enlargement ap-

pears development of what we call the moral nature.
What we name moral feelings become so because in
them .personality determines itself in lines of rec-
titude.

Personal character, though exceedingly complex,
is largely the result of volition manipulating other
psychic elements, and in turn manipulated by them.
As every cognitive state leaves its traces in memory
and in modified mental habitudes, and as every feeling
adds its increment of modification, so every volition
cuts its own figures and works its own results; hence
regret, self-contempt and remorse, and the opposite.
A Commodus, a Cæsar Borgia, or a Marat alongside of
a St. Cecilia, a Francis of Assisi, or a George Washington show how low or how high human character may
fall or rise. The mind is a tablet, on which has been
engraven tokens of all deeds and feelings, good and
bad; and discernment of character is the truthful reading of these fateful hieroglyphs. You are what you
have made yourself to become. Circumstances condition the outer appearance and the superficial display of
the nature; but responsible choice carves out the permanent moral character. Plato, in his Gorgias, pictures Rhadamanthus as finding the soul of the tyrant
" full of the prints and scars of prejudices and wrongs,
which have been stamped there by each action." Lu-
cian, in his Dialogues of the Dead, makes the departed
strip before the judge for examination; and he avers
that burned in upon the breast of every one are found
marks left by the sins of a past life, unseen of mortal
eyes yet visible to divine justice. And our own Tennyson only voices the moral sense of the whole world,

and puts into picturesque form the conclusions of profoundest ethical philosophy, when he sings of the unjust man, that he

> ". . . bears about
> A silent court of justice in his breast,
> Himself the judge and jury, and himself
> The prisoner at the bar, ever condemned."

INDEX.

THE END.

JAMES SULLY'S WORKS.

OUTLINES OF PSYCHOLOGY, with Special Reference to the Theory of Education. A Text-Book for Colleges. By JAMES SULLY, M. A., LL. D., Examiner for the Moral Sciences Tripos in the University of Cambridge, etc. Crown 8vo. Cloth, $3.00.

A book that has been long wanted by all who are engaged in the business of teaching and desire to master its principles. A new and most desirable feature is the educational applications that are made throughout in separate text and type.

TEACHER'S HAND-BOOK OF PSYCHOLOGY. On the Basis of "Outlines of Psychology." Abridged by the author for the use of Teachers, Schools, Reading-Circles, and Students generally. 12mo. 415 pages. Cloth, $1.50.

CONTENTS.—Psychology and Education.—Scope and Method of Psychology.— Mind and Body.—Knowing, Feeling, and Willing.—Mental Development.—Attention.—The Senses : Sense Discrimination, Observation of Things.—Mental Reproduction : Memory, Constructive Imagination.—Abstraction and Conception.—Judging and Reasoning.—The Feelings : Nature of Feeling, The Egoistic and Social Feelings, The Higher Sentiments.—The Will : Voluntary Movement.—Moral Action : Character.—Appendices.

ILLUSIONS : A Psychological Study. 12mo. 372 pages. Cloth, $1.50.

CONTENTS.—The Study of Illusion.—The Classification of Illusions.—Illusions of Perception.—Dreams.—Illusions of Introspection.—Other Quasi-Presentative Illusions ; Errors of Insight, Illusions of Memory, Illusions of Belief.—Results.

"This is not a technical work, but one of wide popular interest, in the principles and results of which every one is concerned, . . . and may be relied upon as representing the present state of knowledge on the important subject to which it is devoted."—*Popular Science Monthly.*

PESSIMISM : A History and a Criticism. Second edition. 8vo. 470 pages and index. Cloth, $4.00.

". . . The necessity of giving new form and point to the discussion has called for this enlarged and essentially new volume. The preface is the brightest possible sketch of recent discussions on the subject and contributions to it. The bibliography which follows is a thorough exhibition of the literature."—*Independent.*

THE HUMAN MIND. A Text-Book of Psychology. 2 vols. 8vo. Cloth, $5.00.

This work is an elaboration of the doctrine set forth in the author's "Outlines of Psychology." Although the mode of arrangement and of treatment is in the main similar, the book is a new and independent publication.

"The exposition is wonderfully clear and readable."—*New York Tribune.*

"It is sufficient to say that by his treatise on the human mind Mr. Sully fully sustains his reputation as a psychologist."—*Nature.*

New York: D. APPLETON & CO., 1, 3, & 5 Bond Street.

ALEXANDER BAIN'S WORKS.

THE SENSES AND THE INTELLECT. 8vo. Cloth, $5.00.

The object of this treatise is to give a full and systematic account of two principal divisions of the science of mind—the senses and the intellect. The value of the third edition of the work is greatly enhanced by an account of the psychology of Aristotle, which has been contributed by Mr. Grote.

THE EMOTIONS AND THE WILL. 8vo. Cloth, $5.00.

The present publication is a sequel to the former one on "The Senses and the Intellect," and completes a systematic exposition of the human mind.

MIND AND BODY. Theories of their Relations. 12mo. Cloth, $1.50.

"A forcible statement of the connection between mind and body, studying their subtile interworkings by the light of the most recent physiological investigations."—*Christian Register.*

EDUCATION AS A SCIENCE. 12mo. Cloth, $1.75.

ON TEACHING ENGLISH. With Detailed Examples and an Inquiry into the Definition of Poetry. 12mo. Cloth, $1.25.

PRACTICAL ESSAYS. 12mo. Cloth, $1.50.

Dr. H. ALLEYNE NICHOLSON'S WORKS.

MANUAL OF ZOÖLOGY, for the Use of Students, with a General Introduction to the Principles of Zoölogy. Second edition. Revised and enlarged, with 243 Woodcuts. 12mo. Cloth, $2.50.

THE ANCIENT LIFE-HISTORY OF THE EARTH. A Comprehensive Outline of the Principles and Leading Facts of Palæontological Science. 12mo. Cloth, $2.00.

"A work by a master in the science who understands the significance of every phenomenon which he records, and knows how to make it reveal its lessons. As regards its value there can scarcely exist two opinions. As a text-book of the historical phase of palæontology it will be indispensable to students, whether specially pursuing geology or biology; and without it no man who aspires even to an outline knowledge of natural science can deem his library complete."—*The Quarterly Journal of Science.*

GEORGE J. ROMANES'S WORKS.

MENTAL EVOLUTION IN MAN: Origin of Human Faculty. One vol., 8vo. Cloth, $3.00.

This work, which follows "Mental Evolution in Animals," by the same author, considers the probable mode of genesis of the human mind from the mind of lower animals, and attempts to show that there is no distinction of kind between man and brute, but, on the contrary, that such distinctions as do exist all admit of being explained, with respect to their evolution, by adequate psychological analysis.

"The vast array of facts, and the sober and solid method of argument employed by Mr. Romanes, will prove, we think, a great gift to knowledge."— *Saturday Review.*

JELLY-FISH, STAR-FISH, AND SEA-URCHINS. Being a Research on Primitive Nervous Systems. 12mo. Cloth, $1.75.

"Although I have throughout kept in view the requirements of a general reader, I have also sought to render the book of service to the working physiologist, by bringing together in one consecutive account all the more important observations and results which have been yielded by this research."—*Extract from Preface.*

"A profound research into the laws of primitive nervous systems conducted by one of the ablest English investigators. Mr. Romanes set up a tent on the beach and examined his beautiful pets for six summers in succession. Such patient and loving work has borne its fruits in a monograph which leaves nothing to be said about jelly-fish, star-fish, and sea-urchins. Every one who has studied the lowest forms of life on the sea-shore admires these objects. But few have any idea of the exquisite delicacy of their structure and their nice adaptation to their place in nature. Mr. Romanes brings out the subtile beauties of the rudimentary organisms, and shows the resemblances they bear to the higher types of creation. His explanations are made more clear by a large number of illustrations."—*New York Journal of Commerce.*

ANIMAL INTELLIGENCE. 12mo. Cloth, $1.75.

"A collection of facts which, though it may merely amuse the unscientific reader, will be a real boon to the student of comparative psychology, for this is the first attempt to present systematically the well-assured results of observation on the mental life of animals."—*Saturday Review.*

MENTAL EVOLUTION IN ANIMALS. With a Posthumous Essay on Instinct, by CHARLES DARWIN. 12mo. Cloth, $2.00.

"Mr. Romanes has followed up his careful enumeration of the facts of 'Animal Intelligence,' contributed to the 'International Scientific Series,' with a work dealing with the successive stages at which the various mental phenomena appear in the scale of life. The present installment displays the same evidence of industry in collecting facts and caution in co-ordinating them by theory as the former."—*The Athenæum.*

" For a still higher order of students, we have a series of 'Classical Writers.' This we can not praise too much."—Westminster Review.

CLASSICAL WRITERS.

EDITED BY JOHN RICHARD GREEN, M. A., LL. D.

16mo, flexible cloth - - - - 60 cents each.

MILTON. By STOPFORD A. BROOKE.

" The life is accompanied by careful synopses of Milton's prose and poetical works, and by scholarly estimates and criticisms of them. Arranged in brief paragraphs, and clothed in a simple and perspicuous style, the volume introduces the pupil directly to the author it describes, and not only familiarizes him with his method of composition, but with his exquisite fancies and lofty conceptions, and enables him to see practically and intelligently what an expressive and sonorous instrument our tongue is in the hands of one of its mightiest masters."—*Harper's Magazine.*

EURIPIDES. By Professor J. P. MAHAFFY.

" A better book on the subject than has previously been written in English. He is scholarly and not pedantic, appreciative and yet just."—*London Academy.*

VERGIL. By Professor H. NETTLESHIP.

" The information is all sound and good, and no such hand-book has before been within the reach of the young student. Any one who wishes to read Vergil intelligently, and not merely to cram so many books of the ' Æneid ' for an examination, should buy Professor Nettleship's scholarly monograph."—*London Athenæum.*

SOPHOCLES. By Professor LEWIS CAMPBELL.

" We can not close without again recommending the little book to all lovers of Sophocles, as an able and eloquent picture of the life and work of one of the greatest dramatic writers the world has ever seen."—*London Athenæum.*

LIVY. By Rev. W. W. CAPES, M. A.

" Well deserves attentive study on many accounts, especially for the variety of its theme and the concise perspicuity of its treatment."—*London Saturday Review.*

DEMOSTHENES. By S. H. BUTCHER, Fellow of University College, Oxford.

" This is an admirable little book. Mr. Butcher has brought his finished scholarship to bear on a difficult but most interesting chapter of Greek literary history ; . . . the primer is as fresh and attractive in form as it is ripe in learning and thorough in method."—*London Academy.*

LANDMARKS OF ENGLISH LITERATURE. By HENRY J. NICOLL. 12mo. Cloth, $1.75.

"The plan adopted in this book has been to deal solely with the very greatest names in the several departments of English literature—with those writers whose works are among the most imperishable glories of Britain, and with whom it is a disgrace for even the busiest to remain unacquainted."
—*From the Preface.*

"The 'Landmarks of English Literature' is a work of exceptional value. It reveals scholarship and high literary ability. Mr. Nicoll has a proper conception of the age in which he lives, and of its requirements in the special line in which he has attempted to work."—*New York Herald.*

"We can warmly recommend this excellent manual."—*St. James's Gazette.*

"Mr. Nicoll is not ambitious, save to state things precisely as they are, to give the common orthodox judgment on great authors and their places in history, and he has brought to his task a mild enthusiasm of style and a conscientiousness of exact statement that can not be overpraised. He writes out of a full mind, and yet he writes on a level with the ordinary intelligence."—*New York Times.*

"It would be hard to find anywhere an example of a more pithy, compact, yet attractive presentation of the real landmarks of the literature than is comprised in this duodecimo of 460 pages."—*New York Home Journal.*

"The work abounds in personal incident and anecdote connected with various authors, which assist the reader in making their acquaintance, and which give to the book a more lively aspect than one of cold criticism."—*New York Observer.*

"A book to be most heartily commended."—*Boston Traveller.*

"It has ample narrative and happy criticisms, and is filled with instructive and entertaining matter admirably presented. It would be hard to suggest improvement in style or arrangement."—*Boston Commonwealth.*

THE DEVELOPMENT OF ENGLISH THOUGHT. The Old English Period. By Brother AZARIAS, Professor of English Literature in Rock Hill College, Maryland. 12mo. Cloth, $1.25.

"In some respects the author has written a text-book superior to any we know now in use. There are few writers so well prepared in what might be termed the technique of Old English history and literature. His chapter on the Kelt and Teuton is admirable."—*New York Times.*

"The work will commend itself to notice for its concise and agreeable style, its logical method, and the philosophic and poetic, as well as historical, treatment of the theme. The author is master of his subject."—*Providence Journal.*

"The author has exhibited great skill in presenting to the reader a clear and correct view of the literature and condition of things in England at that remote age, and the work is one of special interest."—*Boston Post.*

"A valuable text-book."—*Boston Globe.*

"A work of remarkable interest."—*Boston Evening Transcript.*

"A book of genuine literary interest and value."—*Cleveland Herald.*

"Within its covers there is a wealth of erudition, research, and scholarly labor, which places the book beside those of Wright, Spalding, and Craik. The English of the writer is a model for clearness and point."—*Utica Daily Observer.*

"One of the most thorough and best-arranged books on the subject that we have seen."—*Troy Press.*

New York: D. APPLETON & CO., 1, 3, 5 Bond Street.

HERBERT SPENCER'S COMPLETE WORKS.

THE works of the philosopher of Evolution consist of the series under the general title of The Synthetic Philosophy, and of the other volumes named below. No other author has developed the principles of evolution so completely and systematically as Spencer has. In the nine volumes of the Synthetic Philosophy he shows that this great process is constantly going on in the universe as a whole and in all (or nearly all) its details; in the aggregate of stars and nebulæ; in the planetary system; in the earth as an inorganic mass; in each organism, vegetal or animal (Von Baer's law); in the aggregate of organisms throughout geologic time; in the mind; in society; in all products of social activity. He makes practical applications to both private conduct and public affairs —to education, commerce, government, philanthropy, religion, and morals.

Some readers have found the technical terms, which a superficial glance has revealed to them, somewhat of a hindrance. Mr. Spencer's style is, however, remarkable for its clearness of statement and its wealth of illustration. The technical terms, which result from condensing phrases into single words, are required for accuracy, and the reading of a chapter or two takes away all their strangeness.

These books present such convincing arguments that they have won the immediate assent and the enthusiastic adherence of hundreds of thousands of readers to the doctrine of evolution, and they are likely to have even greater influence in the future than they have had in the past.

SYNTHETIC PHILOSOPHY:

First Principles. 1 vol., 12mo. Cloth, $2.00.

    I. The Unknowable.   II. Laws of the Knowable.

The Principles of Biology. 2 vols., 12mo. Cloth, $4.00.

I. The Data of Biology.	IV. Morphological Development.
II. The Inductions of Biology.	V. Physiological Development.
III. The Evolution of Life.	VI. Laws of Multiplication.

The Principles of Psychology. 2 vols., 12mo. Cloth, $4.00.

I. The Data of Psychology.	V. Physical Synthesis.
II. The Inductions of Psychology.	VI. Special Analysis.
	VII. General Analysis.
III. General Synthesis.	VIII. Congruities.
IV. Special Synthesis.	IX. Corollaries.

The Principles of Sociology. 2 vols., 12mo. Cloth, $4.00.

I. The Data of Sociology.	III. The Domestic Relations.
II. The Inductions of Sociology.	IV. Ceremonial Institutions.
	V. Political Institutions.

      VI. Ecclesiastical Institutions.

www.ingramcontent.com/pod-product-compliance
Lightning Source LLC
Chambersburg PA
CBHW030406270326
41926CB00009B/1287